Textbooks in Electrical and Electronic Engineering

1. Introduction to fields and circuits
   GORDON LANCASTER

2. The CD-ROM and optical recording systems
   E.W. WILLIAMS

3. Engineering electromagnetism: physical processes and computation
   P. HAMMOND AND J.K. SYKULSKI

4. Integrated circuit engineering
   L.J. HERBST

5. Electrical circuits and systems: an introduction for engineers and physical scientists
   A.M. HOWATSON

6. Applied numerical modelling for engineers
   DONARD DE COGAN AND ANNE DE COGAN

Applied Numerical Modelling for Engineers

■

Donard de Cogan
Reader in Electronics, University of East Anglia, Norwich

and

Anne de Cogan
Senior Lecturer in Computing, City College, Norwich

OXFORD   NEW YORK   TORONTO
OXFORD UNIVERSITY PRESS
1997

Oxford University Press, Great Clarendon Street, Oxford OX2 6DP
Oxford New York
Athens Auckland Bangkok Bogota Bombay Buenos Aires
Calcutta Cape Town Dar es Salaam Delhi Florence Hong Kong
Istanbul Karachi Kuala Lumpur Madras Madrid Melbourne
Mexico City Nairobi Paris Singapore Taipei Tokyo Toronto Warsaw
and associated companies in
Berlin Ibadan

Oxford is a trade mark of Oxford University Press

Published in the United States
by Oxford University Press Inc., New York
© Donard de Cogan and Anne de Cogan, 1997

A catalogue record for this book is available from the British Library

Library of Congress Cataloging in Publication Data
De Cogan, Donard.
Applied numerical modelling for engineers / Donard de Cogan and Anne de Cogan
(Textbook in electrical and electronic engineering; 6)
Includes bibliographical references and index.
1. Engineering–Mathematical models. I. De Cogan, Anne.
II. Title. III. Series.
TA342.D43 1997   621.3'01'1–dc21   96–49957
ISBN 0 19 856437 6 (p/b)
0 19 856438 4 (h/b)
Typeset by EXPO Holdings, Malaysia

Printed in Great Britain by
Bookcraft (Bath) Ltd,
Midsomer Norton, Avon

Preface

The main function of this book is to act as a tutorial guide which shows how a range of techniques at the physical and mathematical level can be used to convert model descriptions into efficient and reliable simulation algorithms. The book is likely to be useful to second- or third-year undergraduates on electrical, electronic, and general engineering courses and to postgraduates requiring introductory material on these topics. The mathematics is generally at the level required by a second-year student, although it becomes progressively more demanding in the later chapters. We have summarized the mathematical fundamentals in the Appendix for reference.

The introduction outlines the modelling philosophy which we will adopt.

Chapter 2 uses a range of examples to distinguish between continuous and discrete approaches to modelling. Models based on finite difference techniques follow directly from this, and are the subject of Chapter 3.

Chapter 4 moves on to distributed electromagnetic analogues based on the transmission-line-matrix (TLM) method which at one level can be thought of as a subset of finite difference. The techniques of TLM can also be interpreted as a set of rules. Chapter 5 broadens the rule concept to include other rule-based modelling techniques. TLM and some rule based models can also be interpreted in terms of transition probabilities. Chapter 6 discusses a range of modelling techniques based on probability and statistics.

Up to this point the emphasis is on discrete modes involving orthogonal coordinate systems (Cartesian and polar). Such discretizations are not always appropriate in finite-element modelling (Chapter 7) and can be irrelevant in frequency-domain modelling (Chapter 8).

Chapter 9 provides brief outlines of several other important topics which have not found a place elsewhere in a book which is heavily biased towards differential methods in the time domain.

Software listings in BASIC, C, Mathematica, and Maple are included throughout the book where relevant. These have been kept deliberately simple and can be adapted by the interested reader.

Exercises are provided at the end of most chapters. These are intended as mini-projects which can be expanded or adapted as required.

Norwich   D. de C.
April 1997   A. de C.

Contents

5 Rule-based models

6 Probability based models

7 Models involving non-Cartesian meshes

1.1 Modelling as a natural human activity

1.1.1 Man makes models

From earliest times, man has attempted to understand the universe of which he is part. His interest has been partly curiousity, partly a practical concern with controlling his environment. Models have allowed him to make explicit the extent of his knowledge and understanding, to explain phenomena, and to predict behaviour. As deficiencies in concepts become apparent they are superseded by improved or totally revised models. Nowhere is this more clearly shown than in the way in which scientists over the centuries have constantly revised and redrawn models of our own universe.

A successful model leads to an improved understanding of a physical situation, but that alone is rarely enough.

1. The model must be robust. There cannot be situations within the limits of its definition and underlying assumptions that lead to inconsistent results.

2. The model must be predictive. It must be able to anticipate results which can subsequently be verified by experimental observation.

The process of using a model to confirm our observations and to predict future behaviour is known as *simulation*.

1.1.2 The life cycle of a model

The predictive aspect of modelling almost invariably requires the translation of a model into a mathematical formulation. Newton's '*fluxions*' (differential calculus) was developed for this purpose. It proved to be so powerful that it still dominates our way of thinking in many areas. Models such as Newton's laws of motion tend to progress through a series of phases which are often described in terms of the stages of human life (Fig. 1.1).

1.1.3 A model is a model

Good models are efficient mirrors of reality. A modeller will endeavour to employ a set of assumptions which simplify the description of physical phenomena with minimal distortion. Unfortunately, such assumptions tend to limit the range of applicability of the model.

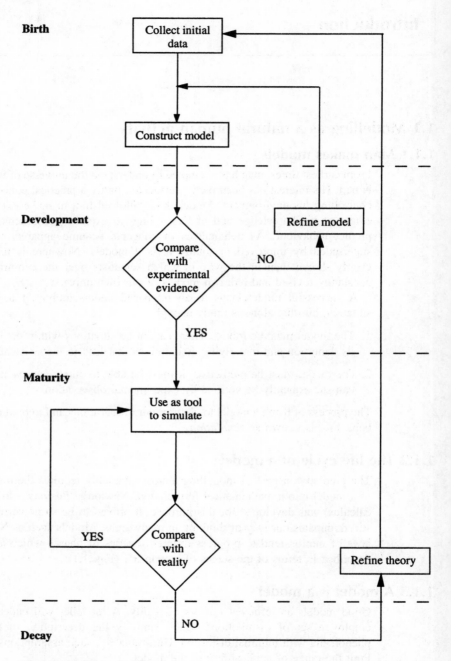

Fig. 1.1 The life cycle of a model.

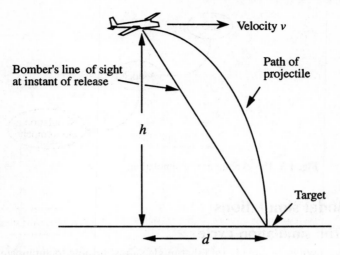

Fig. 1.2 A typical modelling problem in warfare.

To illustrate this, we can consider a first year engineering physics problem, Newton's laws of motion in relation to the trajectory of a free-fall bomb released from an aircraft (Fig. 1.2).

The simple Newtonian equations for motion in a vacuum relate the initial velocity u to the distance covered S or the instantaneous velocity v to the lapsed time t and the acceleration due to gravity g:

$$v = u + gt$$
$$S = ut + \frac{1}{2}gt^2$$
$$v^2 = u^2 + 2gS$$

These equations will probably ensure acceptable hit rates at low altitudes where frictional drag factors are of limited significance. At higher altitudes it will be neccessary to add contributions due to air viscosity, laminar flow around the projectile, gyroscopic motion, and possibly Coriolis forces.

This raises a further problem. The enhanced model is certainly more realistic in physical terms. However it is also much more complicated in terms of its mathematical formulation. Unless simplifications are made in the mathematical analysis, the resulting equations will be almost impossible to solve. The development of digital computers during the period of the Second World War was in large part a response to the need to calculate accurate long-range artillery trajectories.

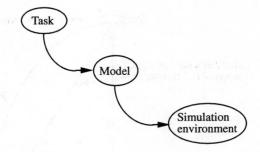

Fig. 1.3 The evolution of a simulation.

1.2 Model simulations

1.2.1 The simulation cycle

We have already said that models must be able to anticipate results which can subsequently be verified by experimental observation. Rather than risk the life of a test pilot, an aeronautic engineer will first model the behaviour of his design before testing a scaled-down version in a wind tunnel. Only when favourable results are consistently obtained will the full-scale version be built. There are other circumstances where simulation may be the only available option: it is highly desirable to know the effects of lightning on modern aircraft. However, quite apart from the safety of those involved and the cost implications, it is not easy to get atmospheric conditions to oblige with thunderstorms to order.

In general, the engineer is first presented with a task (Fig. 1.3). The formal specification may be anything but trivial. At this point there is a need for a clear physical description of the problem together with an indication of the variables involved. One or more models are independently developed according to the scheme in Fig. 1.1. These are then integrated and tested within a simulation environment. The equivalent development cycle is shown in Fig. 1.4.

Many simulation environments are available as general-purpose computer software packages (Maple, Mathematica, Matlab, Stripes, Pafec), reducing the drudge of manual calculation and greatly extending computability. There has also been much work on the development of efficient user interfaces which both present the information in a comprehensible form and protect the core software from uninformed adaptation.

Caution is needed in using any simulation process. Unless we clearly understand the limits of applicability of the environment, it is likely that we may attempt to exceed the operating range of the method.

As demands on them increase, simulation environments may be amended, modified or even expanded. Sooner or later however we will approach or even

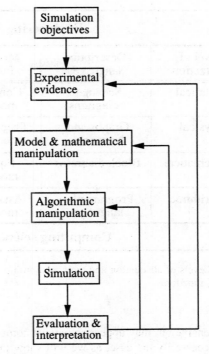

Fig. 1.4 A typical simulation sequence.

exceed the bounds of computability. Parkinson's law says that work expands to fill the time available for its completion. In this respect it could be restated as:

Models expand to the limits of the power of computers which are available for simulation.

1.2.2 Two types of simulation

In a physical problem there may well be interaction between several of the process variables.

1. In a *what-if* approach, the user will vary one or more parameters at a time and to investigate the overall effect on a design.

2. The *ensemble* approach seeks to consider all possible responses to all known variables.

The *what-if* approach tends to ignore the influences of statistical fluctuations which are inherent in real systems. The *ensemble* approach is generally more

Engineering			
Level of abstraction	Description languages	Modelling function	VLSI chip equivalent
Technical	Schematic diagrams	Component modelling	Transistor level description
Physical	Graphs and Networks	Phenomenon modelling	Logic gate description
Mathematical	Block diagrams	Equation modelling	Functional description
Algorithmic	Programming languages	Assignment modelling	System description
Computing science			

Fig. 1.5 Levels of abstraction and their relationship to description language and modelling function.

comprehensive but may appear somewhat computer intensive. In practice this is not neccessarily the case, as we will show in Chapter 6.

1.3 The systems approach to modelling and simulation

The concept of systems design in engineering is intuitive, but the enormous increases in the complexity of electronic circuits over the last fifteen years have forced engineers to apply formal methods to systems thinking. Most modern electronic systems are designed at a functional level using a system description language (SDL).

In this context we can usefully relate the different descriptions to the levels of abstraction which are currently used in model development. The diagram shown in Fig. 1.5 is adapted from Lorentz [1.1].

Within the Lorentz scheme a modelling task may be initiated at either the physical or the technical level of abstraction. The task may be described in many ways, but formal description languages assist with analysis and with the communication of concepts. The 'physical' and 'mathematical' levels are used to develop the model and the 'mathematical' and 'algorithmic' levels are important in the development of the simulation environment. There is currently some movement towards reducing the dependence on the mathematical level and proceeding from the physical level directly to the algorithmic level.

In this chapter we have presented the basis of an effective approach to modelling. In the next chapter we look at some examples and distinguish those which are amenable to an analytical treatment from those which require a numerical approach. Most of the succeeding chapters are concerned with different numerical techniques which can be used in modelling, however, the concepts which have been outlined in this chapter should be applied regardless of the solution strategy which is adopted.

References

1.1 F. Lorentz, *A multiformalism modelling approach using together bond graphs, networks and block diagrams,* Proceedings of Eurotherm-36, Poitiers, 1994.

Models: continuous and discrete

Since the time of Isaac Newton mathematicians have developed a substantial armoury of analytical techniques based on calculus which can be used to model physical processes. In this chapter we start by considering a range of techniques which can be used either singly or together to construct models. We then move on to the simulation of physical problems using these models. We discuss the information we need about initial and boundary conditions to predict future system behaviour. A knowledge of such additional factors may be available from experimental observations or may indeed constitute an integral part of the model itself.

Analogue descriptions

It is not always essential to develop a model in terms of its own variables. Reasons of conceptual visualization or computational efficiency some-times make it more convenient to describe a problem by means of one or more analogues which can then be translated into a mathematical description. For example, electrical concepts of current, voltage, and resistance can be used to model the flow of water in a pipe. They can also be used to model mechanical systems such as springs and shock absorbers. Analogue descriptions will form an important part of this and later chapters.

Discrete descriptions

In many cases it is easier to treat problems as discrete rather than continuous descriptions of a real system. In particular, it is frequently possible to use discrete methods in situations where analytical solutions would be impractical. It may be that within the level of required accuracy such a method is adequate. The level of pollution in a lake, for example, is sampled at certain points only and ecologists may use this data to express an opinion about the overall condition of the lake. This chapter closes with several examples of discretised approximations to real problems. The techniques for manipulating such models will be discussed in later chapters.

2.1 Continuous models and analytic solutions

2.1.1 From practical observation to mathematical description direct

The pendulum and simple harmonic motion

The Dutch scientist, Christiaan Huygens is credited with the development of the pendulum clock which revolutionized timekeeping. In common with other scientists he undertook experiments to determine the effect of the mass of a pendulum and its length. Observations at that time and since have led to a simple and well-known second-order differential equation for the case where the instantaneous angle to the vertical is kept small:

$$\frac{d^2\theta}{dt^2} = -\omega^2\theta \tag{2.1}$$

(θ is the angle to the vertical and $\omega = 2\pi f$ where f is the frequency).

The effect of pendulum length and mass under these circmstances were investigated and it was found that the frequency was proportional to the inverse root of the length and was independent of mass. A more comprehensive mechanical analysis of a pendulum would not restrict θ to small values and would include the effects of air resistance and perhaps friction at the fulcrum point.

Nuclear decay and other first order processes

The process of radioactive decay is another instance where it is possible to proceed direct from observation to formulation. The experimentalist might start with a quantity of Si^{31} which decays to P^{31} with the emission of beta particles. Observations of the intensity of beta emission might be taken as a measure of the concentration of the isotope and it would not be long before the decay could be seen as exponential. One confirmation of this would be the observation that the half-life was constant and independent of time.

In any system which involves exponential decay, the time which it takes for the concentration $N(t)$ to reach half the original value N_0 is given by

$$N(t_{1/2}) = N_0 e^{-\frac{t_{half}}{c}}$$

where C is a constant, but since

$$N(t_{1/2}) = \frac{N_0}{2}$$

$$\therefore \quad \frac{1}{2} = e^{-\frac{t_{half}}{c}}$$

$$\log_e 2 = \frac{t_{half}}{c}$$

or

$$t_{half} = c \, / \log_e 2.$$

This decay behaviour can be formulated by another simple differential equation

$$-\frac{dN}{dt} = k_d N. \tag{2.2}$$

That is, the rate of reduction of species N at any instant depends only on the quantity present at that time. There are many reactions in chemistry where the rate of decay is proportional to the first power of the concentration and these are said to obey first order kinetics. In general terms the quantity of reactant or product is given as a concentration and is shown in square brackets as below.

> Concentrations of a chemical in solution could be quoted as molecules per litre, but this might create some problems. However, since equal quantities of atoms of different materials are related through their atomic weights, it follows that equal quantities of molecules can be related through mass. The *mole* is defined as the mass of a material which contains one Avogadro number of atoms/molecules. Hydrogen has an atomic mass of 1 and chlorine has an atomic mass of 35.5. Thus 3.65 gm of hydrogen chloride (HCl) in one litre of water (hydrochloric acid) represents $[HCl] = 0.1$ moles/litre.

Chemical process with heat generation

The reaction of carbon and oxygen is known to be exothermic: i.e. heat is generated.

$$C + O_2 \rightarrow CO_2$$

It is possible to show by means of experimental observations that the rate of removal of carbon or oxygen or the rate of production of carbon dioxide is proportional to the product of the two concentrations.

$$-\frac{d[C]}{dt} = -\frac{d[O_2]}{dt} = \frac{d[CO_2]}{dt} = k_f \, [C] \, [O_2]. \tag{2.3}$$

The constant of proportionality, k_f, is called the forward rate constant

The process in eqn (2.3) is said to be second order because the sum of powers of the concentrations is two.

Although considerable heat is generated the process is not spontaneous. The reactants must be heated in order to overcome an initial barrier. This is called the activation energy. It is shown in Fig. 2.1 along with the heat of formation, which is the net energy given out during the reaction.

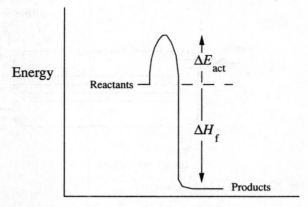

Fig. 2.1 Representation of energy changes during a chemical reaction.

The process which is described by eqn (2.3) does not contain temperature and yet it is well known that the rate of reaction approximately doubles for every 10 °C rise in temperature. In fact, the temperature dependence is hidden within the rate constant and measurements of the dependence of k_f on temperature reveal that

$$k_f = k_{f0} \, e^{\frac{-\Delta E_{act}}{kT}} \qquad (2.4)$$

where, k is Boltzman's constant and T is the absolute temperature. The quantity ΔE_{act} is the activation energy for the reaction.

Since the the energy level of the products lies below that of the reactants this implies that the heat of formation (ΔH_f) is generated by the process. It is expressed as a normalized quantity which represents the energy (joules) emitted per molecule formed. From the viewpoint of a chemical engineer an exothermic process can present considerable problems in terms or safety and/ or economy. If the generated heat cannot be removed in an efficient manner then the temperature rise increases the rate of reaction, increases the heat output and ... an explosion could ensue. On the other hand, if heat can be removed effectively then it may be possible to use it to do useful work and thereby reduce primary energy costs. This can be achieved by means of a heat exchanger which consists of a structure where the reacting mixture and coolant fluid are mechanically separated by a thin metal membrane.

Steady-state heat transfer by conduction

Continuing this example, which is an important aspect of chemical engineering, we could consider a chemical plant with heat exchangers comprising a series of tubes through which the reacting species was allowed to pass at a constant flow rate in order to allow heat to be transferred to a cooling

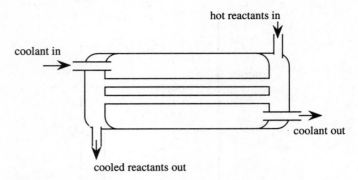

Fig. 2.2 A typical heat exchanger.

liquid flowing at a constant rate in the opposite direction on the other side of the tubes (Fig. 2.2). If this process were allowed to continue over a long time then the system would approach an equilibrium or steady state. This condition would prevail when the rate of heat transfer at any point as well as the temperature distribution throughout the system did not change with time.

It might perhaps be easier to visualize a furnace with a uniform lining which is operating at 1200 °C within a constant temperature ambient. A series of thermocouples could be used to measure the temperature at different positions x within the lining. Such observations would indicate that temperature gradient was constant. This could be expressed by a simple differential equation:

$$\frac{\partial^2 T}{\partial x^2} = 0. \tag{2.5}$$

This could be extended to further dimensions. Figure 2.3 shows such an example of steady-state heat flow where the governing differential equation is

$$\frac{\partial^2 T}{\partial x^2} + \frac{\partial^2 T}{\partial y^2} = 0. \tag{2.6}$$

This is called *Laplace's equation* and is an example of a special family of mathematical descriptions called *elliptic* equations.

If the conducting medium shown in Fig. 2.3 were originally at 18 °C and were brought into intimate contact with 100 °C constant heat sources at the start of an experiment, then there would be a transient period as it heated up. Ultimately, a stage would be reached where there were a set of time-independent temperature contours (isotherms) across the surface.

Steady-state heat transfer by free convection

Convection is a process of heat transfer which frequently involves exchange between materials of different phase (solid/liquid, solid/gas, liquid/gas). We

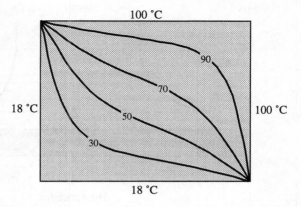

Fig. 2.3 Steady-state heating of a material with annotated contours.

might think that a model for a central heating radiator located against a wall in a large hall is simple, but it has exercised the minds of some of the greatest mathematicians. There is no difficulty in writing down an expression for the local heat flux:

$$q'' = h(T_s - T_\infty) \tag{2.7}$$

where h is the local convection coefficient. However, as in the case of reaction kinetics, this coefficient conceals a vast spectrum of physics. This problem involves heat transfer, mass transfer, and fluid flow.

All the equations of fluid mechanics can be brought to bear on this problem. We can use the Euler equations [2.1] or the Navier–Stokes equation [2.2] if the viscosity of the medium must be considered.

Steady-state heat transfer by forced convection

If we re-examine the problem of the heat exchanger then the situation is somewhat simpler than the case of free convection. Flow is due to an external force rather than to bouyancy effects within the fluid.

The fluid responding to the external pressure may exhibit either laminar flow or turbulent flow. In the latter case the motion is irregular and there are velocity fluctuations. However in the case of laminar flow it is possible to identify streamlines, along which material moves. The velocity is zero at the surface, and increases to a constant value u_∞ as we move away, normal to the surface. This is shown in Fig. 2.4 along with the temperature variation that might be expected in the steady state. Above all, it is possible to use well-known differential equations to express the behaviour in this relatively simple problem.

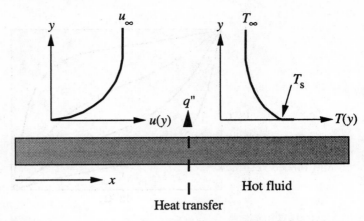

Heat transfer

Fig. 2.4 Fluid velocity and steady-state temperature distribution as a function of distance (y) from the heating surface in forced convection.

These examples have largely covered cases where the experimental observations have involved a one-way progression in time and/or space which may ultimately reach a steady-state condition. Even the pendulum will be subject to frictional forces which eventually result in it coming to rest. However the subject of periodic processes is vast and it is therefore worth devoting some more attention to it within the context of this chapter.

Modelling the generation of musical sounds

A single vibration in a string may be described using the expression for simple harmonic motion: the displacement $y(t)$ of the string from its central position is given by the equation

$$\frac{d^2 y(t)}{dt^2} = -\omega^2 y(t) \tag{2.8}$$

where $\omega = 2\pi f$ (f is the frequency).

However reality is not as simple as that otherwise we would not be able to distinguish between different musical instruments. It is now known that any sound signal contains a spectum of frequencies. In the absence of oscilloscopes, researchers in the last century used a series of resonators to determine the relative magnitudes of these harmonics and it soon became obvious that the components of frequency in a vibrating string could be represented by a Fourier series (see Section 8.3.1):

$$f(x) = \frac{a_0}{2} + \sum_{n=1}^{\infty} [a_n \cos(nx) + b_n \sin(nx)] \tag{2.9}$$

where $a_0 = \displaystyle\int\limits_{-\pi}^{\pi} f(x)\,\mathrm{d}x, \quad a_n = \displaystyle\int\limits_{-\pi}^{\pi} f(x)\cos(nx)\,\mathrm{d}x, \quad b_n = \displaystyle\int\limits_{-\pi}^{\pi} f(x)\sin(nx)\,\mathrm{d}x.$

For example, the action of a bow on a violin string might be represented by a saw-tooth waveform ($f(x) = x$ in the range $-\pi$ to π) given by

$$f(x) = 2\left[\sin(x) - \frac{\sin(2x)}{2} + \frac{\sin(2x)}{2} - \ldots\right].$$

Damped and forced resonances

It is traditional that marching soldiers break step when crossing a bridge. This is supposed to be due to the fact that the pound of the step may be at the resonant frequency of the structure and could lead ultimately to destruction. The basic idea can be expressed as:

$$\frac{\mathrm{d}^2 y}{\mathrm{d}t^2} = -\omega^2 y + A\cos\omega t. \tag{2.10}$$

The $A\cos\omega t$ term represents an excitation at a frequency which is identical with the resonant frequency of the system. The effect is shown in Fig. 2.5.

It is interesting to see how this increase in the amplitude is entirely due to resonant build-up. Figure 2.6 shows the results for a similar equation with an excitation which is a twice the resonant frequency of the system. It is quite clear that there is no build-up as in Fig. 2.5.

It is doubtful whether the effects of marching would lead to the failure of a bridge which is after all a damped resonator (i.e. even if excited, it will

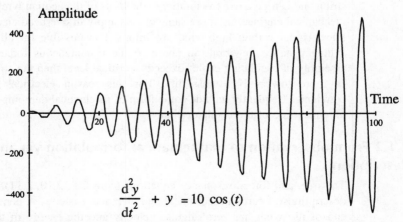

Fig. 2.5 Amplitude vs. time solution of an equation of the form of eqn (2.10).

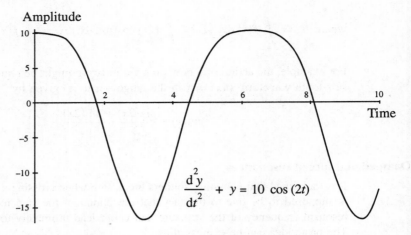

Fig. 2.6 Amplitude vs. time solution for an excitation at frequency $2f$.

eventually revert to zero amplitude oscillations). A situation where both the resonator and the forcing agency are both damped can be expressed by

$$\frac{d^2y}{dt^2} + 2h\frac{dy}{dt} = -(\omega^2 + h^2)y + A\,e^{-ht}\cos\omega t. \tag{2.11}$$

It can be seen in Fig. 2.7 that even with small friction the amplitude does not tend towards infinity. This result is somewhat similar to a large pendulum on a clock which is driven from a small external source.

We conclude this section by considering a large-scale example of what has just been outlined. A wind blowing past wires can act like a violin bow on a string and can give rise to vibrations (the basis of the aeolian lyre). In terms of mechanical engineering these same effects apply in overhead electrical power lines. Under certain high-wind conditions it is possible for all cables over a long distance to vibrate in phase. If the instantaneous forces due to the assembly of oscillating cables exceeds a critical level then the collapse of one or more towers is a possiblity. For this reason electrical transmission companies frequently include some form of mechanical damping on the wires on either side of the insulator chains.

2.1.2 From observation to mathematical formulation via an analogue description

The free, and forced resonance equations (eqns 2.8, 2.10, 2.11) are excellent descriptions of certain mechanical and electrical systems. A pendulum at one end of its swing has zero angular velocity and the energy of the system is stored as potential energy. At the lowest point of its swing potential energy is

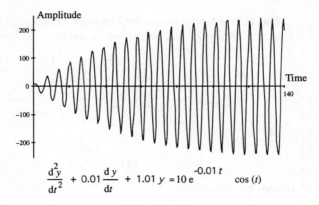

$$\frac{d^2y}{dt^2} + 0.01\frac{dy}{dt} + 1.01\,y = 10\,e^{-0.01\,t}\,\cos(t)$$

Fig. 2.7 Amplitude–time plot for an equation of the form of eqn (2.11).

zero while the kinetic energy is a maximum. It can therefore be described as a system which stores energy in one form and periodically undergoes conversion between the two forms. A spring is another example of exactly the same thing. In electrical engineering a capacitor is a store of electrostatic energy ($E_C = \frac{1}{2}CV^2$). An inductor is a store of magnetic energy, which is dependent on the instantaneous current ($E_L = \frac{1}{2}LI^2$). If an inductor and capacitor are connected together then it is possible to have energy exchange between them and this leads to a resonance which can be expressed by means of eqn 2.8. The forced resonance condition can be simulated by driving a spring or an LC network with an external source of the same frequency.

Once it has been accepted that an adequate mathematical description of many physical phenomena is possible, we can then approach modelling from the other side: start with the nearest physical analogue and then use the mathematics that has been formulated for describing problems of that type. Some examples are discussed below:

Example 2.1 Damped resonance

Figure 2.8a shows a parallel connection of inductor, resistor and capacitor.

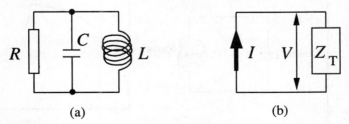

(a)   (b)

Fig. 2.8 (a) Parallel connection of inductor, capacitor, and resistor which can be used to model a damped resonance. (b) Black box representation of the same circuit.

This circuit can be represented by three impedances in parallel, so that a total impedance, Z_T can be defined in terms of the sum of reciprocals.

$$\frac{1}{Z_T} = \frac{1}{R} + \frac{1}{Z_C} + \frac{1}{Z_L} \tag{2.11}$$

where $\quad Z_L = j\omega L \quad$ and $\quad Z_C = \dfrac{1}{j\omega C}$ $(j = \sqrt{-1})$.

Thus,

$$Z_T = \frac{j\omega LR}{R(1 - \omega^2 LC) + j\omega L}.$$

Resonance occurs when

$$\omega^2 LC = \quad (\text{or } f_{res} = \frac{1}{2\pi}\sqrt{\frac{1}{LC}})$$

and under these conditions the total impedance of the circuit is R. In more general terms a circuit can be said to be in resonance when the impedance is real. From this arrangement we can use Ohm's law, $(V = I\, Z_T)$ to determine the current response of the circuit (Fig. 2.8b) due to an excitation voltage of any frequency and amplitude.

Example 2.2 Viscoelastic behaviour
Mechanical analogues can be extremely useful. A series combination of spring and dashpot (shock absorber) can be used to model a viscoelastic (Maxwell) liquid where strain rises linearly with time. A parallel combination of the same components models a viscoelastic solid where the strain is proportional to $(1 - e^{-ct})$, for some constant c. A viscoelastic solid which experiences flow as a result of stress can be modelled by a three component combination and this is shown in Fig. 2.9.

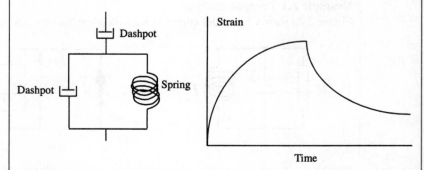

Fig. 2.9 A three component mechanical analogue of irreversible flow in a viscoelastic solid. The graph shows typical strain/time response for constant stress.

Example 2.3 Transient heat conduction

Heat propagation in solids is another case where a very effective description can be obtained using a circuit analogue. For one-dimensional conduction an RC ladder network as shown in Fig. 2.10 is adequate.

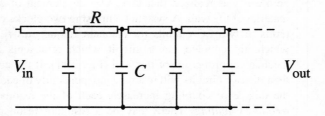

Fig. 2.10 An RC ladder network of the type used to model transient heat conduction or the transmission of signals on early telegraph cables.

This circuit can be used as an analogue for the two laws of diffusion which were proposed by Fick and which can be expressed in one dimension as:

$$\phi = -D\frac{\partial N}{\partial x}$$

That is the flux ϕ (rate of flow of N per unit area) is proportional to the concentration gradient. The minus sign implies that the flow is always down the gradient. Flux and gradient are related through the diffusion constant, D:

$$D\frac{\partial^2 N(x,t)}{\partial x^2} = \frac{\partial N(x,t)}{\partial t}.$$

The circuit in Fig. 2.10 can be used to treat both of these diffusion laws analogously. If we take the tops of two adjacent capacitors we see that they are joined by a resistor. The difference in voltage across a resistor gives rise to an effective electric field ($-dV/dx$). There is a flow of current through the resistor which is proportional to the field (the basis of Ohm's law).

Let us assume that V_{in} was switched on at $t = 0$. An analysis of the spatial and temporal transient behaviour would lead to:

$$\frac{\partial^2 V(x,t)}{\partial x^2} = RC\frac{\partial V(x,t)}{\partial t} \tag{2.12}$$

which is identical with Fick's second law of diffusion with $D = (RC)^{-1}$.

Example 2.4 Coupled resonant systems

Analogue descriptions are especially useful as the problem being modelled becomes more complex. The story is told that the early manufacturers of pendulum clocks experienced a curious phenomenon. So long as one clock was in position on a wall it kept good time, but if other pendulum clocks were placed on the same wall, then none were accurate. It was some time before it was realized that there was an element of coupling through the material of the wall. A system comprising two clocks could be expressed in terms of two LC circuits (of resonant frequencies f_1 and f_2 respectively) which are coupled via a circuit which represents an analogue of the mechanical properties of the wall (Fig. 2.11). If the coupling is very small then the two circuits will resonate independently of each other. However, as the degree of coupling increases, each of the resonators will experience *frequency pulling*. There may be a resultant resonance with side bands which represent the individual contributions, but since each circuit has some influence on the other, neither will be at their original resonance point.

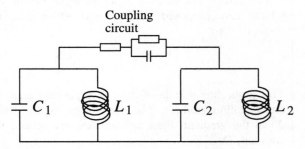

Fig. 2.11 Coupled resonant circuits.

Example 2.5 Properties of dielectrics

Electrical circuits are frequently used to provide analogues for other electrical circuits. One such case is the use of additional resistors and capacitors to simulate dielectric materials in circuits. Dielectrics are a class of insulating materials which are important in the electrical and electronic industry. They include the solids that separate the plates of a capacitor or the fluids which provide insulation in high-voltage transformers.

Two separated electrodes will store opposite value charges. The amount of charge is proportional to the voltage difference between the electrodes. The area and the separation between the electrodes have an influence on the constant of proportionality. But so also does the material (dielectric) which separates capacitor plates. A dielectric increases the charge/volt that can be stored on the plates and this is expressed in terms of a parameter called the relative permittivity ε_r which influences the free space capacitance, C_0, so that $C = \varepsilon_r \, C_0$. Under the influence of an applied AC voltage (V_{max} $\sin(\omega t)$), an ideal capacitor should pass a current which is 90° out of phase. In real capacitors the phase angle is less than 90° by an amount δ. The resolution of vectors shows that a component, $I \sin(\delta)$ is in phase with the applied voltage and this gives rise to power dissipation. The parameter, δ, or more strictly, $\tan(\delta)$ is called the loss angle and is the measure of the quality of a dielectric (small δ means a good dielectric).

This effect can be modelled by a parallel connection of resistor R_p and capacitor C_p. The overall effect can be expressed mathematically as:

$$\varepsilon_r = \varepsilon' - j\,\varepsilon'' \tag{2.13}$$

The phase angle, $\frac{\varepsilon''}{\varepsilon'}$ of this complex number is in fact the loss angle δ.

Within this formulation $C_p = \varepsilon' \, C_0$ and $R_p = \dfrac{1}{\varepsilon'' \omega C_0}$.

It can be seen that the power dissipation becomes greater as the frequency of the signal is increased. Dielectrics which are used in high frequency applications must have very small loss angle in order to minimize losses.

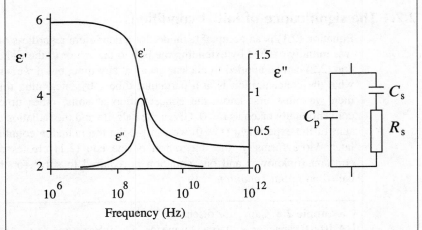

Fig. 2.12 Typical relaxation spectrum of a transformer oil and the equivalent circuit which provides a good model for this behaviour.

Comprehensive measurements on a large range of dielectrics indicate that both ε' and ε'' are functions of frequency. For many, the value of ε' decreases with frequency while the value of ε'' goes through a maximum.

These are called relaxation spectra and an example is shown in Fig. 2.12 along with the circuit analogue which gives the best approximation to this behaviour. In physical terms this means that the individual dipoles within the dielectric attempt to follow the changing potential as it goes from positive to negative and back again. At low frequencies this does not present a problem, but as the frequency is increased, the ability to respond decreases (just like a bass loudspeaker being asked to transmit a treble sound). As more dipoles assume a neutral position, the real part of permittivity (ε') decreases. Within this roll-off process there is a point of maximum power dissipation, the peak in the ε'' curve. The essential feature of all of this is the fact that we can use an electrical network which will express the behaviour of the circuit and this feature has wider applications which will become obvious later.

2.2 Simulation using analytical treatments

Once observations have led to an analytical description in the form of a differential equation the next step is to try to predict behaviour by simulation. However some further information, in the form of initial and boundary conditions, is required before that can be achieved.

2.2.1 The significance of initial conditions

Equation (2.1) is an acceptable model for a pendulum regardless of whether it was initially excited by extending the bob to the left or to the right. Similarly, eqn (2.2) can be applied to nuclear decay at any time, but if we wish to know what the concentration is at a particular time τ by integrating this equation, then we must first know the concentration at some other time, which is conventionally taken as $t = 0$. Given the half life and the radiation count at the start of the experiment ($t = 0$) we can predict the radiation count at any time later. More formally, the integration of the eqn (2.1) creates an arbitrary constant (unknown) and one equation is required to solve for this. This is called an *initial condition*.

Example 2.6 Capacitor discharge
A 10 μF capacitor is charged by means of a 10 V battery. At $t = 0$ the battery is removed and replaced by a 10 000 Ω resistor which allows the capacitor to discharge. It is required to find the voltage across the battery after 0.1 s.
This problem can be treated by a modification of eqn (2.2):

$$-\frac{dV}{dt} = \frac{1}{RC}V \tag{2.14}$$

Integrating this and knowing that $V = 10$ V at $t = 0$ we get a general predictive solution

$$V(t) = 10\,e^{\frac{-t}{RC}}. \qquad (2.15)$$

Since $RC = (10^4)(10^{-5}) = 10^{-1}$ s, $V(0.1 \text{ s}) = 3.678$ volts.

The principles of algebraic equations apply equally well here: if there are n unknowns then there must be n equations for their solution.

Example 2.7 Chemical reaction with heat generation
We now revisit the combustion of carbon in oxygen (eqn (2.3)) where it is obvious that the prediction of concentrations at any time requires information about the concentrations of both species at some fixed time. In order to be more widely applicable, the equation can be written in a general form:

$$\frac{d[P]}{dt} = -\frac{d[R_1]}{dt} = -\frac{d[R_2]}{dt} = k\,[R_1]\,[R_2] \qquad (2.16)$$

where $[P]$, $[R_1]$, and $[R_2]$ are the concentrations of product, reactant 1 and reactant 2 respectively.

If at $t = 0$ the reactants had initial concentrations $a = [R_1]_{t=0}$ and $b = [R_2]_{t=0}$ and if a quantity of product α was produced after time t then eqn (2.16) could be rearranged as

$$\frac{d\alpha}{(a - \alpha)(b - \alpha)} = k\,dt. \qquad (2.17)$$

Integration of this gives the concentration of product at any time t:

$$\alpha(t) = \frac{ab\left[1 - e^{-kt(a-b)}\right]}{\left[a - be^{-kt(a-b)}\right]}. \qquad (2.18)$$

This applies so long as $a \neq b$ and the dependence of α on time for $a = 5.9999$, $b = 6$, $k = 0.01$ is plotted in Fig. 2.13.

If the initial concentrations a and b were identical then it is easier to use the differential equation:

$$-\frac{d[R]}{dt} = k\,[R]^2.$$

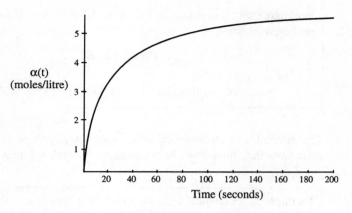

Fig. 2.13 Plot of the concentration of products in a second-order chemical reaction where the initial concentrations are $a = 5.999$ moles/litre and $b = 6$ moles/litre and the rate constant is $b = 0.01$ litres/mole-sec.

This yields an instantaneous value of reactant according to the equation:

$$\frac{1}{[R]_t} - \frac{1}{[R]_0} = kt.$$

In either event if the initial concentrations were quoted we would know the concentration of species which was produced within a unit interval of time. We would also know the quantity of heat which is generated when a unit concentration of species is produced. Thus we could estimate the heat liberated within an interval of time.

The data in Fig. 2.13 could be used to estimate the increase in concentration of product between 100 and 110 seconds (0.074 moles/litre). Now if the heat of reaction $\Delta H_f = -20$ kJ/mole then this means that 1.48 kJ of heat was liberated per litre of solution (or 1.48 J/ml) during 10 seconds centred on 105 seconds. It so happens that 4.18 J raises the temperature of 1 ml of water by 1 °C. So, if the reaction was undertaken in well stirred aqueous solution, then every millilitre of liquid would experience a temperature rise of 0.354 °C during this time interval.

Note Although this is quite a modest temperature rise, the temperature increase earlier in the reaction would have been much greater. You may have noticed that this analysis assumed the efficient removal of heat so as to ensure that the rate constant k remained constant throughout (see eqn (2.4)).

Example 2.8 Motion of an electron in a magnetic field [2.3]
An electron moving in a uniform magnetic field perpendicular to its direction of motion experiences a force which can be given by the simultaneous differential equations:

$$m\frac{d^2y(t)}{dt^2} = -\frac{He}{c}\frac{dx(t)}{dt}$$

$$m\frac{d^2x(t)}{dt^2} = -\frac{He}{c}\frac{dy(t)}{dt} \tag{2.19}$$

H is the field, c is the velocity of light, m and e are the mass and charge of the electron.

We can define sensible initial conditions: at time $t = 0$ the electron is at $x = 0$, $y = 0$, but is moving at a velocity u in the x direction. There is no component of velocity in the y direction.

It is then possible to solve (2.19) to get:

$$x(t) = \frac{ucm}{He}\sin(\frac{He}{cm}t)$$

$$y(t) = \frac{ucm}{He}[1 - \cos(\frac{He}{cm}t)]. \tag{2.20}$$

Note This is an extremely important set of equations which amongst other things predict the motion of an electron spot within a TV tube.

2.2.2 The significance of boundary conditions

If the string of a musical instrument were plucked at time $t = 0$ then this would provide an initial condition. It has been shown above that this on its own would not provide sufficient information for a full solution. The disturbance caused by the pluck would propagate along the string and after some time would arrive at a termination. If we knew the length of the string and the velocity of propagation, then we would be able to derive another piece of information, namely that at time $t = L/v$ the wave could not escape and thus would have to have zero amplitude. This additional piece of information would come from a knowledge of one or more of the boundaries and is thus termed a boundary condition.

Wave motion within a constrained space

The equation for wave motion can also be expressed in terms of spatial derivatives:

$$\frac{d^2y(x)}{dx^2} = -k^2y(x) \tag{2.21}$$

where k is the wave number ($= 2\pi/\lambda$). Solutions can be of the same form as previously. The most general solution is

$$y(x) = A \sin(kx) + B \cos(kx).$$

If the wave is constrained to exist only between $x = 0$ and $x = L$, then this provides sufficient boundary conditions.

$$\text{At } x = 0$$
$$0 = A \sin(0) + B \cos(0) = 0 + B.$$

Therefore $B = 0$.

$$\text{At } x = L$$
$$0 = A \sin(kL).$$

Either $A = 0$ (the wave has zero amplitude) or there are other conditions which operate. In fact $\sin(kL)$ is zero if $kL = 0$, π, 2π, 3π, ... This means that eqn (2.21) is satisfied for any wave with $\lambda = 2L/n$ ($n = 1, 2, 3, \ldots$), which brings us back towards Fourier's theorem again.

Example 2.9 Diffusion from a sphere
This example is concerned with a sphere where material diffuses radially into its surrounds. It demonstrates how Fourier techniques can be used to predict concentration as a function of time and position subject to initial and boundary conditions. This might apply to a medical implant which delivers a drug to the body over a period of time.

For this problem eqn (2.12) has to be written in spherical polar coordinates:

$$\frac{\partial^2 C(r,t)}{\partial r^2} + \frac{2}{r}\frac{\partial C(r,t)}{\partial r} = \frac{1}{D}\frac{\partial C(r,t)}{\partial t}. \tag{2.22}$$

We can apply a transformation $U(r, t) = C(r, t)r$ so that eqn (2.22) becomes

$$\frac{\partial^2 U(r,t)}{\partial r^2} = \frac{1}{D}\frac{\partial U(r,t)}{\partial t}. \tag{2.23}$$

The solution of this equation requires boundary conditions which might reasonably be:

at $t = 0$ $C(x, 0) = C_0$ for all $x < a$ (where a is the radius of the sphere);

for $t \geq 0$ $C(x, t) \to 0$ as $t \to$ for all $x < a$;

for $t \geq 0$ $C(a,t) = 0$ (i.e. species at the surface is removed).

When these are applied to a general solution given as a Fourier series (eqn (2.9)) it it possible to determine the values of the constants [2.4]. In this case the solution at any time t and position r in a sphere of radius a is:

$$C(r,t) = -2\,C_0 \sum_{n=1}^{\infty} (-1)^n \left[\frac{a}{n\pi r}\right] \sin\left[\frac{n\pi r}{a}\right] \mathrm{e}^{-\frac{(n\pi)^2 Dt}{a^2}}. \tag{2.24}$$

On the basis of this model the pharmaceutical company might advise doctors on how often it is necessary to replace a slow drug release implant.

Example 2.10   The field inside a semiconductor punch-through diode
An unconnected semiconductor diode is a system which is only in thermodynamic equilibrium because the flux of carriers due to diffusion is balanced by the electrostatic field. This gives rise to regions of material where there is a fixed charge and regions which are neutral. If a single electron were to travel from one side of a diode without causing an appreciable disturbance it would experience regions where Laplace's equation operated and regions where Poisson's equation operated. Poisson's equation applies when there is a charge distribution within the region. This, in one-dimensional space, is:

$$\frac{\partial^2 V}{\partial x^2} = -\frac{\rho}{\varepsilon_r \varepsilon_0} \tag{2.25}$$

The equation can be integrated with respect to the distance x, subject to boundary conditions. One of the benefits of this type of analysis is a prediction of the equilibrium width of the region where carriers have diffused to leave a charged (depleted) region [2.5].

A punch-through diode is a special structure where the depleted region extends throughout a distance L as shown in Fig. 2.14. The charge density is $-eN_d$, where N_d is the carrier concentration. Equation (2.25) applies and is subject to the boundary condition that at $x = 0$, $V = 0$. The active length is L. At any applied voltage less than a critical value the device does not carry current. However, when $V \geq V_{pt}$, the device is said to be in punch-through and a significant current flows. The value of this critical voltage is given by:

$$V_{pt} = \frac{1}{2} \frac{eN_d L^2}{\varepsilon_r \varepsilon_0}. \tag{2.26}$$

Note The local variation of potential has an important technological significance. Since the potential at any point inside the structure is

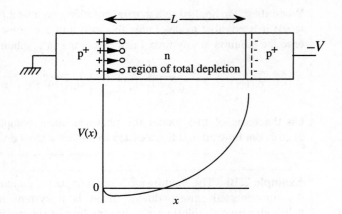

Fig. 2.14 A p^+np^+ diode biased just above punch-through showing the injection of holes on the left. The graph of $V(x)$ vs. x immediately below indicates the local potential which the holes will experience as they move from left to right.

proportional to x^2 then the power dissipation (given by $IV(x)$) is also proportional to x^2. In order to reduce the overall operating temperature of this type of device when used for electrical overload protection it is important to arrange the heat sinking to be on the high potential side [2.6].

The cases which have been discussed above have indicated the importance of boundaries. In fact boundaries are classified in a formal way as follows:

1. A situation where the dependent variable (e.g. $V(x)$) is defined on the boundary is termed a *Dirichlet* problem.

2. When the derivative of the dependent variable $(dV(x)/dx)$ is given on the boundary then it is called a *Neumann* problem.

3. *Mixed* boundary problems have the variable prescribed over part of the boundary and the derivative prescribed over the remainder.

2.2.3 Problems with real boundaries

The major problem with real boundaries is that they are not necessarily well behaved in a mathematical sense. The natural level of complexity may make the type of analytical solution we have described almost impossible to compute.

Example 2.11 Sound propagation in shallow underwater (sea) environments: a case of non-ideal boundaries

From an analytical viewpoint, eqns (2.8) and (2.21) can be combined to give an alternative wave equation which is the generic form of hyperbolic equations. The propagation of sound waves can be given by a modification of eqn (2.21).

$$\frac{d^2 y(x)}{dx^2} = \frac{1}{v^2} \frac{d^2 y(t)}{dt^2}. \tag{2.27}$$

This provides an excellent description of the propagation of longitudinal waves where the velocity $v = \sqrt{\kappa/\rho}$ where κ is the bulk modulus for the medium and ρ is the density.

However accurate simulations of this problem present considerable difficulties associated with boundary descriptions. The water surface has waves which can vary in amplitude and wavelength. Signals transmitted at different times will behave differently since the nature of the surface is continuously changing. At low sound frequencies (long wavelength) it is possible to treat the surface as a uniform reflecting layer. At higher frequencies the sound will experience diffraction as if from an optical grating.

Added to this, there are inhomogeneities. On a bright day, the water temperature near the sea surface rises which results in a decrease in density. This local effect will cause refraction so that the influence of the actual surface boundary may be reduced or possibly completely removed.

Finally, the sea bed is also a non-ideal boundary. It is inhomogeneous and the reflection coefficient for sound depends on the material and the state of compaction. It also depends on the signal frequency and on angle of incidence. In summary, what appears to be a simple problem can become computationally impossible as approximations to reality are attempted by the addition of different features to the initial model.

Similar problems exist in electrostatics. It is known that the Laplace equation in its most general form

$$\nabla^2 V = \frac{\partial^2 V}{\partial x^2} + \frac{\partial^2 V}{\partial y^2} + \frac{\partial^2 V}{\partial z^2} = 0 \tag{2.28}$$

provides a good model for the electrostatic potential, $V(x, y, z)$, in a region of space which is free of electrical charges. An accurate simulation may be very important in the design and positioning of a lightning protection system for a new building. For this purpose the assumption that a thunder cloud can be treated as some distant electrode gives reasonable analytical results.

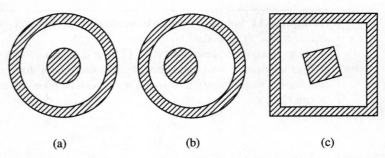

Fig. 2.15 Electrode geometries that present increasing levels of difficulty for analytical solution of the fields in the intervening space.

The Laplace equation can also be applied to cylindrical and spherical geometries, (subject to a transformation into the appropriate coordinate system). The value of potential can be obtained analytically for any point between the electrodes in a coaxial system of conductors as shown in Fig. 2.15(a). The assymetry in Fig. 2.15(b) makes the mathematics becomes much more difficult, but a solution is possible [2.7].

A real engineering problem might involve a rectangular cylindrical conductor inside a bigger rectangular duct with non-coincidence of diagonals, as in Fig. 2.15(c). This arrangement might be found in a safety-critical environment (such as coal mines or oil rigs) with a need to estimate the maximum potential difference between electrodes to ensure that an electrical discharge will never occur. To solve this problem we must consider using numerical methods.

2.3 The case for numerical simulations

There is a long and honourable history of using numerical and alternative methods for obtaining solutions where it may be difficult or indeed impossible to obtain an analytical solution. It is not unknown for chemistry students to plot an integral on squared graph paper and then to count squares by means of accurate weight measurements. If the weight of a known area of graph paper is measured then the integral (area under the curve) can be evaluated. Simpson's rule and the trapezium rule (see Appendix) are examples of piecewise numerical integration.

Simplifications and approximations such as Taylor expansions can be used. A Maclaurin series provides an estimate of functions which can be expanded to any required level of accuracy. For example

$$\sin(x) = x - \frac{x^3}{3!} + \frac{x^5}{5!} - \dots . \tag{2.29}$$

If x is small, for example 0.1745 radians (10°) then $\sin(x) = 0.1736$ so that the first term in the Maclaurin expansion is satisfactory as a rough approximation.

2.3.1 A word of caution

It has been said that chemists make pure substances, physicists make accurate measurements and that physical chemists make accurate measurements on pure substances. So what of the engineer?

The answer is obvious: engineers make things happen. An engineer must carry the sum total of knowledge within and beyond a field of expertise and must understand the properties of materials and their interactions in a real (i.e. impure) environment. Above all, an engineer must develop an intuition about what is inherently right and must exercise caution, where appropriate. Such intuitive caution applies here.

In addition to the considerations about modelling and simulation in the previous chapter we could also add some general cautions which are important for the engineer/modeller. A problem analysis may result in a mathematical model which contains a quadratic equation. Such equations have two roots which may have different signs. Does this mean that our model can have negative temperatures or pressures? There are other problems with trigonometrical functions: $\tan(x)$ will have one value but $\cosh^{-1}(x)$ can have two values and $\tan^{-1}(x)$ has an infinite number of satisfactory values. A calculation may involve taking the difference between two very large numbers. The result may depend more on the numerical precision limitations of the computer than on anything that has physical meaning. These cautions are even more important in the discrete modelling approach which is considered next.

2.4 The discrete approach

If a single half cycle from Fig. 2.7 were magnified then it would probably look more like the picture shown in Fig. 2.16. The discretized image can arise from a variety of sources: the accuracy limits of the plotting routines, the quantization of the software that was used in the conversion between computer image file formats and at the finest level (not apparent here) from the discretization of the CRT display. In spite of these effects, the discretisation has not adversely affected the perception of the contents of Fig. 2.7. Indeed images in newsprint and pictures on television screens are also discretized and provided they are viewed from a reasonable distance, there is no appreciable loss of information.

The underlying message is:

There is little point in having a mathematical simulation that is accurate to 100 decimal places if the best display or plotting equipment is only accurate to ± 1%.

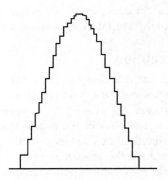

Fig. 2.16 Magnified image of a single half cycle from Fig. 2.7.

Thus instead of defining a derivative as:

$$\lim_{\Delta x \to 0} \frac{f(x + \Delta x) - f(x)}{\Delta x} \qquad (2.30)$$

we can define it as $f\frac{(x+\Delta x)-f(x)}{\Delta x}$ where Δx is sufficiently small as not to introduce appreciable error. This *finite-difference* approach is the major subject of the next chapter.

2.4.1 Time and space discretizations

A physical problem such as the bending of a supported beam under its own weight could be broken down into a series of interconnected regions in space (spatial discretizations) where the analysis can be undertaken on the assumption that within any individual region (node) the properties may be treated as uniform. If the beam is narrow, then it will be possible to treat it in one dimension. However the apprentice modeller might still pose the question:

How will I know when I have got the discretization right?

The answer to this is quite simple: *try it and compare it.* If the error, when compared with an analytical treatment (or against some benchmark) is acceptable, then the discretization is reasonable. It may be that there is no available analytical treatment in which case it is a good idea to start with a coarse discretization and to refine it until a level is reached where there is no change in the result within the limits of required accuracy. The method is generally satisfactory, but it is worth remembering that the final result could be the product of a consistent and scalable error or could be the convergence to the appropriate result of an essentially different problem. If possible other, perhaps experimental, confirmation should be sought.

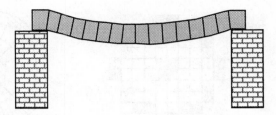

Fig. 2.17 A discretized model of a supported beam where the use of a finer two-dimensional mesh might be advisable.

If the thickness of the beam that was being modelled was appreciable, then it might be advisable to use a two-dimensional mesh as the nodes in the upper part of the beam will be under compression and those in the lower part will be in tension (Fig. 2.17). In other words the stress varies across the thickness of the beam and the discretization must take account of this.

Similar arguments can also be used for time discretization. So long as no property which is a significant part of a model changes appreciably during the time interval then there will be litttle loss of accuracy. A model of heat conduction through a safety fire door might produce acceptable results with a time step of 10 minutes, whereas a simulation of laser cutting of metal might require a time step of the order of microseconds. There are situations where materials of different thermal properties are in close contact and the optimum choice of time step may be a problem. The simulation of thermal transients of tube and shell heat exchangers (see Fig. 2.2) fall within this catagory. This will be discussed later, but is due to the fact that the metallic components have a very rapid thermal response whilst the fluid can have a very slow thermal response. If the time step and spatial discretizations are appropriate to the fluid only, then the predicted behaviour of heat flow in the metal will be at least wrong and could result in numerical instability. At the other end of the spectrum discretizations more appropriate to the metal would result in over definition of the fluid region in space and time. This is computationally inefficient as well as time consuming.

2.4.2 Rectangular versus cylindrical coordinates

In many situations a rectangular Cartesian coordinate system may be suitable for numerical simulation. The same technique can sometimes be applied without too much loss of accuracy to cylindrical problems. Thus it may be possible use the approximation shown in Fig. 2.18a to treat the electrode system of Fig. 2.15a. If this is not possible, then a polar mesh (Fig. 2.18b) must be used.

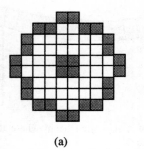 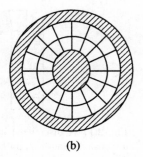

(a)   (b)

Fig. 2.18 (a) Rectangular Cartesian approximation to Fig. 2.15a. (b) Polar coordinate representation of Fig. 2.15a.

2.4.3 Uniform and non-uniform discretizations

At first sight it might seem a waste of effort to use a non-uniform mesh but this is not always so. The simulation of heat conduction from the centre of a cylinder is a case in point. The conventional method might involve a discretization scheme which had a constant radial distance Δr (Fig. 2.19a). If this is used then the physical volumes of successive nodes (counting from the centre) will increase. In certain simulations there may be a benefit in maintaining a constant discretized thermal capacitance throughout the problem (Fig. 2.19b).

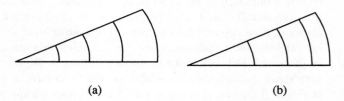

(a)   (b)

Fig. 2.19 A single segment from (a) a uniform; (b) a non-uniform cylindrical polar mesh scheme. The latter scheme has nodes of equal area.

Modelling of carrier transport in semiconductor devices may demand the use of a non-uniform discretization. If we take a diode where the doping concentration follows an expression of the form:

$$C(x) = c_1 e^{-c_2 x^2}. \tag{2.31}$$

The variation of doping concentration with distance can range over many orders of magnitude as can be seen in the plot of carrier concentration versus distance for a thyristor shown in Fig. 2.20. It is often very difficult to achieve a stable simulation using a constant dimension mesh. Within the device the greatest changes are happening when the concentration gradients are largest. A logarithmic scheme for spatial discretization will alleviate the situation.

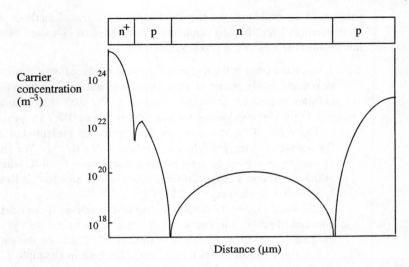

Fig. 2.20 A thyristor structure showing the local variation of carrier concentration, defined as donor concentration–acceptor concentration.

What has been said about variable meshes does not only apply to Cartesian meshes. It will be seen in Section 7.2 that the finite element scheme (which may involve triangular or quadrilateral meshing) will increase the density of the discretization around regions of fine geometry or where significant events are likely to occur.

References

2.1 D.E. Rutherford, *Fluid dynamics*, Oliver & Boyd 1959 p. 19.
2.2 D.E. Rutherford, *Fluid dynamics*, Oliver & Boyd 1959 pp. 194–221.
2.3 A.C. Grove, *An introduction to the Laplace and Z-transform*, Prentice–Hall, 1991, p. 16.
2.4 J. Crank, N.R. Mc Farlane, J.C. Newby, G.D. Paterson, and J.B. Pedley, *Diffusion processes in environmental systems*, Macmillan 1981 pp. 45–52.
2.5 D. de Cogan, *Solid state devices*, Macmillan, London 1987, p. 106.
2.6 D. de Cogan and S. A. John, Failure modes in punch-through diodes, *J. Phys. D. (Applied Physics)*, **18**, 497–505, (1985).
2.7 D.K. Cheng, *Field and wave electromagnetics*, Addison Wesley 1989 p. 169.

Examples, exercises, and projects

In the examples and exercises which are presented below it may in some cases be possible to obtain an algebraic solution. These can of course be solved using one of the many available maths packages. In other cases the complete solution of the problem may require the use of numerical methods which will be covered in later chapters. Our overall objective at this stage is to gain an insight into the modelling process.

2.1 It was mentioned in the text that if the heat produced in a chemical reaction is not efficiently removed then the temperature rise will cause the rate constant to increase according to eqn (2.4). We shall attempt to investigate this further by considering the reaction in Example 2.7 in greater detail.

The value of the rate constant, $k_f = 0.01$ was measured at 27 °C and the activation energy $-\Delta E_{act}$ was 0.56 eV (9×10^{-20} J). We should now be able to determine k_{f0} remembering that T must be in Kelvin (27 °C = 300 Kelvin) and k is the Boltzman constant (1.38×10^{-23} J/Kelvin). Plot k_f against T in the range 0–200 °C.

Chemical engineers frequently apply the rule-of-thumb that the rate constant doubles for every 10 °C rise in temperature. Test this by plotting $k_f = 0.01$ at 27 °C, $k_f = 0.02$ at 37 °C etc. on the same graph.

Finally, let us consider a case where the heat in Example 2.7 was not efficiently removed so that the temperature throughout the reaction rose by 0.2 °C/s. Using $a = 5.9$, $b = 6$ start the reaction at 18 °C and plot $\alpha(t)$ vs. t in eqn (2.18) applying the appropriate value of k_f at every time step and compare the results with Fig. 2.13.

2.2 Consider a layer of ice which is formed on a pond in winter when the air temperature is T_a and the water at the interface with the ice is at 0 °C. If an increment of thickness, dh is to be formed on the ice then a quantity of heat, dQ representing the latent heat of freezing must be transported through the existing layer. Show that the thickness of the ice at any time for any air temperature is given by:

$$h = \sqrt{\frac{2(0 - T_a)\,t}{k_T L_f\, \rho}}$$

where k_T, L_f, and ρ are the thermal conductivity of ice, latent heat of freezing of water, and density of water respectively. Note that there is a similar transport process in the oxidation of silicon (Problem 3.6) where $\sqrt{t}$ behaviour is also observed.

2.3 The pressure of the air at any height in a uniform (ideal) air column can be given by Perrin's law. The change in pressure due to an increment dh to an existing column of height h is :

$$P - (P + dP) = mg$$

where m is the mass of the gas $= (M/V)dh$. In this case M is the molecular weight and V is the volume occupied by one mole of gas.

Use Boyle's law ($PV = RT$) to show that the pressure at height h is related to the ground pressure by:

$$P = P_0 \, e^{\frac{Mgh}{RT}}.$$

As an extension, it might be worth considering how this formula might have been different if the gas had not been ideal and it had been necessary to use van der Waal's equation.

$$\left(P + \frac{a}{V^2}\right)(V - b) = RT.$$

Would it be possible to derive an analytical solution, or would it be better to use a numerical approach?

2.4 A ship in distress fires a flare in order to summon assistance. It is known that the maximum range is achieved when the launch angle is 45°. In the situation where there is a wind the range will be greater if the flare is launched with the wind.

Assuming a launch velocity, V_f and a wind-speed V_w with a frictional force in both directions which is proportional to the square of the velocity in that direction, which direction of launch results in maximum height of the flare, which will consequently be visible at the greatest distance?

If the skipper of the ship was more concerned that the flare should spend as long as possible above his ship, how should he adjust the launch angle?

2.5 A glass beaker (200 mm long and 60 mm diameter) is filled with water and an excitation on the upper lip using a violin bow leads to a resonance. Consider the situation immediately after the beaker is filled with heavily aerated water. Initially the water is milky white, but the white cloud soon starts to lift so that after 5 seconds it occupies only the upper half of the beaker. After 10 seconds it occupies only the upper quarter of the beaker, and so on. How does the resonant frequency of this system vary as a function of time?

2.6 A solar heating panel allows ultraviolet light to pass through a layer of glass before being converted into heat. The glass reflects all infrared radiation, but the heat can conduct out through the glass. Thus the amount of ultraviolet light which passes through the glass is given by the Beer–Lambert law:

$$I_{in} = I_0 \, e^{-kx}$$

where I_0 is the radiation per second per unit area, x is the thickness of the glass layer and k is the glass absorption coefficient. Obviously, the thinner the glass, the greater the transfer of ultraviolet light. However the heat lost per second per unit area is given by:

$$I_{out} = \frac{\Delta T}{k_T x}$$

where k_T is the thermal conductivity of the glass and ΔT is the temperature difference across the glass. In this case, the thicker the glass, the lower the heat loss. If the conversion was 100% efficient then the next heat gain would lead to a rise in temperature per second:

δT = net energy gain per second /(thermal capacitance)

So that

$$\delta T = (I_{in} - I_{out}) / (m\, C_p)$$

where m and C_p are the mass and specific heat of the ultraviolet absorber which achieves the conversion.

Derive an expression for the net heat transfer and for the dependence of x_{min} (the optimum thickness for maximum heat transfer) on the material parameters of the system.

2.7 A clay used in the manufacture of ceramic pipes is formed into a long cylinder and placed in a firing kiln which heats by means of radiation from a uniform source in the roof. Because of a mechanical fault the rollers which turn the cylinder during firing are not working so that the pipe receives radiation through an angle π only. Assuming that the incident radiation (watts cm^{-2}) is completely absorbed within 1 mm of the surface of the 10 cm diameter pipe (wall thickness = 2 cm, density ρ, specific heat C_p and thermal conductivity k_T) prepare a description of the heat input. We can return to this problem later using the numerical techniques in Chapters 3, 4, 5, and 7 with typical values for china clay to obtain temperature profiles as a function of time at different positions within the pipe.

2.8 A sphere of radius r is initially at 1227 °C in free space. The nature of the surface is such that it can be treated as a black-body radiator so that Stefan's law of radiation applies (heat per second per area = σT^4 where the temperature is in Kelvin). Using the parameters (density, specific heat, and thermal conductivity) develop a symbolic model for thermal conduction within the sphere based on a one-dimensional approach. We will return to this problem again in Chapter 3.

The concepts of Problems 2.7 and 2.8 could be combined to consider a sphere of similar consistency to the earth heated during 12 hours (equinox) and then allowed to cool radiatively during 12 hours. This could be approached by considering only a small portion of the sphere with heating which is a function of the angle of the sun. If this thermal cycling model is run for long enough then it should be possible to obtain an indication of the mean temperature at that position.

The finite-difference approach to discrete modelling

3

In the last chapter we showed how a range of problems can be modelled using mathematical formulations. We then indicated how these could be solved either analytically or numerically. We will now go further with the concept of numerical solution. A model based on a differential equation may be amenable to an analytical solution, but in many cases it may simply be more convenient to solve it numerically.

Since a derivative of the function $y = f(x)$ is defined as

$$\frac{dy}{dx} = \lim_{\Delta x \to 0} \frac{f(x + \Delta x) - f(x)}{\Delta x} \tag{3.1}$$

let us consider a problem represented by the differential equation:

$$\frac{dy}{dx} = A$$

(Where A is a constant). We can then use an approximation by considering that the left and right hand sides of eqn (3.1) are equal even when Δx is a small non-zero quantity. Replacing the derivative with its algebraic approximation we then get:

$$\frac{y(x + \Delta x) - y(x)}{\Delta x} = A. \tag{3.2}$$

So, if $y(x)$ is known we can find a solution for $y(x + \Delta x)$, and we can then proceed in a stepwise manner to approximate

$$y(x_1) \quad \text{where } x_1 = x_0 + \Delta x,$$
$$y(x_2) \quad \text{where } x_2 = x_1 + \Delta x = x_0 + 2\Delta x$$

and so on.

This is called the *single step* method and is described in Section 3.1.1. It is the basis of the finite difference technique, but it is not always completely satisfactory. Most of Section 3.1 is devoted to increasingly sophisticated refinements of the method which overcome a tendency to generate significant errors and instabilities in specific cases. In Sections 3.2 and 3.3 the finite-difference treatment is extended to the numerical solution of second-order differential equations which are functions of one or more variables. We will discuss the accuracy and stability of different methods, some of which are quite complex, bearing in mind the modeller's main objective of obtaining optimum results with minimum resources.

3.1 Techniques for first-order initial-value problems

We start by considering various techniques for solving problems which are described by first order differential equations. This technique is applicable to all such equations provided a value of y is known for some initial value of x.
We therefore have

$$y' = \frac{dy}{dx} = f(x, y) \tag{3.3}$$

where $y(x_0) = y_0$.

In the following sections we will express the finite interval (Δx above) as 'h'. This is consistent with the presentations in many other texts.

3.1.1 Single-step methods

In the simple formulations below, new values are calculated in terms of the previous value only. In Section 3.1.2 we move on to consider multi-step techniques which use several previous values in the approximation of each new value.

The Euler–Cauchy method

Using the basic Taylor series

$$y(x + h) = y(x) + h\,y'(x) + \frac{h^2}{2!}y''(x) + \ldots \tag{3.4}$$

We can rewrite eqn 3.3 as:

$$y(x + h) = y(x) + h f(x, y) + \frac{h^2}{2!}f'(x, y) + \ldots \tag{3.5}$$

If h is sufficiently small that h^2 and higher terms can be ignored then we have

$$y(x + h) = y(x) + h f(x, y). \tag{3.6}$$

This is the Euler–Cauchy method. The accuracy of the results depend on the curvature of the function described by the differential equation. In the case of a straight line ($y = mx + c$) there are no higher order terms and the outcome is exact.

It is not difficult to identify equations for which the Euler–Cauchy method is unsatisfactory, for example:

$$\frac{dy}{dx} = x^2, \text{ where } y(0) = 0.$$

The exact solution to this equation is $y = \frac{x^3}{3}$.

The Euler–Cauchy solution for this equation is $y(x_{n+1}) = y(x_n) + h x_n$ and if we take $h = 0.5$ then we can demonstrate the first few steps:

$n = 1 \quad x_1 = 0.5 \qquad (x_1)^2 = 0.25 \quad h(x_1)^2 = 0.125 \qquad y(x_1) = y(x_0) + 0.125$

$n = 2 \quad x_2 = 1.0 \qquad (x_2)^2 = 1.0 \quad\; h(x_2)^2 = 0.5 \qquad\;\; y(x_2) = y(x_1) + 0.125$

$n = 3 \quad x_3 = 1.5 \qquad . \qquad\quad . \qquad\quad . \qquad\quad . \qquad\quad . \qquad\quad . \qquad .$

In Table 3.1 we compare the analytical results with the Euler–Cauchy approximations, using $h = 0.5$. It is clear that a large error has been introduced. While the size of the error could be reduced by using a smaller value of h, the inaccuracy is intrinsically due to the curvature of the function.

The improved Euler or predictor–corrector method

The accuracy of the Euler–Cauchy method can be improved by using eqn (3.6) to provide a first estimate of the value at y_{n+1}, given the value at y_n

$$y^*_{n+1} = y_n + h f(x_n, y_n). \tag{3.7}$$

We then attempt to correct the estimate by averaging the values.

$$y_{n+1} = y_n + \frac{h}{2}[f(x_n, y_n) + f(x_{n+1}, y^*_{n+1})]. \tag{3.8}$$

For convenience we use what are called auxiliaries, k_1 and k_2. The sequence of steps in the new algorithm is:

$$x_{n+1} = x_n + h$$
$$k_1 = h f(x_n, y_n)$$
$$k_2 = h f(x_{n+1}, y_n + k_1)$$
$$y_{n+1} = y_n + \frac{1}{2}[k_1 + k_2].$$

We can apply this predictor–corrector method to the same differential equation that was solved using the Euler–Cauchy method and the first few steps with $h = 0.5$ are given below:

$x_1 = 0.5 \quad k_1 = h(x_0)^2 \quad k_2 = h(x_1)^2 \quad y(x_1) = y(x_0) + \dfrac{0 + 0.125}{2} = 0.0625$

$x_2 = 1.0 \quad k_1 = h(x_1)^2 \quad k_2 = h(x_2)^2 \quad y(x_2) = y(x_1) + \dfrac{0.0625 + 0.3125}{2}$

$$= 0.3850$$

$x_3 = 1.5 \quad k_1 = h(x_2)^2 \quad k_2 = h(x_3)^2 \quad . \quad . \quad .$

Table 3.1 Solutions of $\frac{dy}{dx} = x^2$ where $y(0) = 0$ (with $h = 0.5$)

n	x_n	y_n (analytical)	y_n (Cauchy–Euler)	y_n (improved Euler)	y_n (Runge–Kutta)
0	0.0	0.0	0.0	0.0	0.0
1	0.5	0.042	0.125	0.0625	0.042
2	1.0	0.333	0.625	0.3850	0.333
3	1.5	1.125	1.75	1.1975	1.125
4	2.0	2.666	3.75	2.760	2.666
5	2.5	5.208	6.875	5.325	5.208
6	3.0	9.000	11.375	9.135	9.000

The actual outcomes, shown in Table 3.1, demonstrate a slightly better agreement with the analytical result over some of the range of x.

The Runge–Kutta method

A technique due to C.D.T. Runge and W. Kutta is very widely used and provides solutions of first-order differential equations to a very high level of accuracy. This is largely because it is a fourth-order method; that is, the terms in the Taylor series are not truncated until h^5 and beyond. This method involves the computation of four auxiliary variables, k_1, k_2, k_3, and k_4. The sequence in the algorithm is defined as follows:

$$x_{n+1} = x_n + h$$
$$k_1 = h f(x_n, y_n)$$
$$k_2 = h f(x_n + h/2, y_n + k_1/2)$$
$$k_3 = h f(x_n + h/2, y_n + k_2/2)$$
$$k_4 = h f(x_n + h, y_n + k_3)$$
$$y_{n+1} = y_n + \frac{1}{6}[k_1 + 2k_2 + 2k_3 + k_4].$$

For many problems there is a closed form expression that removes the need for an iterative algebraic solution. We could take for example the differential equation:

$$\frac{dy}{dx} = x \text{ where } y(0) = 0$$

with $h = 0.2$ we could write:

$$x_{n+1} = x_n + 0.2$$
$$k_1 = 0.2x_n$$
$$k_2 = 0.2\,(x_n + 0.1)$$
$$k_3 = 0.2\,(x_n + 0.1)$$
$$k_4 = 0.2\,(x_n + 0.2).$$

Thus

$$y_{n+1} = y_n + \frac{0.2}{6} [6x_n + 0.6] = y_n + 0.2x_n + 0.02.$$

The same approach can be applied to the differential equation that was solved using the Euler–Cauchy and improved Euler methods. In this case the algebra is more protracted and access to software such as Mathematica, Maple, or MathCAD is helpful. The Maple input for this is:

```
interface (prettyprint = Ø):
h:= 0.5;
k1:= h*(Xn^2);
k2:= h*((Xn + h/2)^2;
k3:= h*((Xn + h/2)^2;
k4:= h*((Xn + h)^2);
Yn1:= simplify(evalf(Yn + (k1 + 2*k2 + 2*k3 + k4)/6));
```

and the output is

```
h := .5
k1 := .5*Xn^2
k2 := .5*(Xn + .25)^2
k3 := .5*(Xn + .25)^2
k4 := .5*(Xn + .5)^2
Yn1 := Yn + .5*Xn^2 + .25*Xn + .041666
```

That is we have:

$$y_{n+1} = y_n + 0.5(x_n)^2 + 0.25x_n + 0.041666$$

The results for the Runge–Kutta method (calculated iteratively) are also given in Table 3.1 and are in exact agreement with the analytical results within the numerical limits of the presented data.

The Runge–Kutta technique can be applied in a wide range of modelling problems and the following is a typical example.

Example 3.1 Optimizing a process in chemical engineering

For the purposes of this example we consider a set of chemical reactions which yield product, D. This is formed by the reaction of starting materials A and B. However, the process is not straightforward, but involves an intermediate, C. The process can be expressed by the following reactions

$$A + B \ \overline{\quad\quad} r_1 \longrightarrow C$$

$$C \xrightarrow{\ r_2\ } A + B$$
$$3C \xrightarrow{\ r_3\ } D$$

The reaction rate constants r_1, r_2, and r_3 are dependent on temperature according to the Arrhenius equations

$$r_1 = r_{10}\, e^{-\frac{\Delta E_1}{kT}}$$
$$r_2 = r_{20}\, e^{-\frac{\Delta E_2}{kT}}$$
$$r_3 = r_{30}\, e^{-\frac{\Delta E_3}{kT}}$$

In each case the values for the zero rate constant (e.g. r_{10}) and barrier height (e.g. ΔE_1) have been well characterized. On the basis that r_2 is significant, the objective for the chemical engineer is to choose a temperature which maximizes the rate of production of D. The treatment, which is outlined below indicates how the concentration of the product [D] and intermediate [C] can be calculated. The next stage would be to repeat the calculations over a range of temperatures and identify a value which optimized d[D]/dt.

The entire process can be expressed in terms of two simultaneous first-order differential equation where the initial values $[A]_0$, $[B]_0$ are given and $[C]_0$ and $[D]_0$ are assumed to be zero:

$$\frac{d[C]}{dt} = r_1[A][B] - r_2[C]$$
$$\frac{d[D]}{dt} = r_3[C]^3.$$

A Runge–Kutta routine can be used to solve for $[C]_{n+1}$ from the first equation and this can then be used to solve for $[D]_{n+1}$ in the second equation. Closed-form expressions may be possible, but most users are content to use iterative code. Perhaps the most important aspect of this analysis is the correct choice of the discretization parameter h which in this case represents intervals of time. If there are significant differences in the rate constants, then h must be chosen so as to give accurate results for the fastest process. An examples of the use of Runge–Kutta routines in the analysis of oscillating chemical reactions can be found in reference [3.1].

3.1.2 Multi-step methods

The techniques which have been outlined up to now have used only values obtained at the previous step to calculate a result. In this section we consider methods which use values from several previous steps.

Adams–Bashford method

If we start with a general initial value problem as expressed in eqn (3.3) then we can integrate in the range x_n to $x_{n+1}(x_{n+1} = x_n + h)$:

$$y(x_{n+1}) - y(x_n) = \int_{x_n}^{x_{n+1}} y'(x)\mathrm{d}x = \int_{x_n}^{x_{n+1}} f(x, y(x))\mathrm{d}x \qquad (3.9)$$

The function f can be replaced by an interpolation polynomial of third degree, $p_3(x)$, treated with a backward difference formula. This procedure is outlined in the Appendix.

Eventually we get:

$$p_3(x) = f_n + r\nabla f_n, + \frac{1}{2}r(r+1)\nabla^2 f_n + \frac{1}{6}r(r+1)(r+2)\nabla^3 f_n + \ldots \quad (3.10)$$

where $r = \dfrac{x - x_n}{h}$, and $\nabla f_n = f_n - f_{n-1}, \nabla^2 f_n = f_n - 2f_{n-1} + f_{n-2}$, etc.

Now, the integral can be restated as:

$$\int_{x_n}^{x_{n+1}} p_3(x)\,\mathrm{d}x = h\left(f_n + \frac{1}{2}\nabla f_n, + \frac{5}{12}\nabla^2 f_n + \frac{3}{8}\nabla^3 f_n\right). \qquad (3.11)$$

Substitution into eqn (3.11) and collection of terms yields:

$$y_{n+1} = y_n + \frac{h}{24}(55f_n - 59f_{n-1} + 37f_{n-2} - 9f_{n-3}). \qquad (3.12)$$

Adams–Moulton method

A further refinement of the Adams–Bashford method is the Adams–Moulton method, which simply treats the expression in eqn (3.12) as an intermediate value, y_{n+1}^*, in a predictor scheme. Thus we can write

$$f_{n+1}^* = f(x_{n+1}, Y_{n+1}^*).$$

The corrector phase is then completed by replacing f_n with f_{n+1} in eqn (3.10) and using $r = (x - x_{n+1})/h$ to get:

$$f_{n+1} + r\nabla f_{n+1} + \frac{1}{2}r(r+1)\nabla^2 f_{n+1} + \frac{1}{6}r(r+1)(r+2)\nabla^3 f_{n+1} + \ldots.$$

This is integrated as in eqn (3.11) and gives

$$y_{n+1} = y_n + \frac{h}{24}(9f_{n+1}^* + 19f_n - 5f_{n-1} + f_{n-2}). \qquad (3.13)$$

This form of the predictor–corrector approach is particularly useful since the difference, $|y_n - y_n|$ can give a measure of the level of error. If this difference is large, then it is an indication that the discretization h should be reduced.

Both the Adams–Bashford and Adams–Moulton methods yield excellent results. They do have one significant drawback: they are not self-starting. Initial values for f_0, f_1, f_2 and f_3 are needed, and it may be necessary to use an alternative method such as Runge–Kutta to obtain these. However, once started, they are faster than Runge–Kutta as there are less values to be calculated at every step.

3.2 Finite difference techniques for second-order initial-value problems

Many of the principles which have been discussed in Section 3.1 are applicable to second order problems. However we now have to deal with a general equation of the form:

$$y'' = \frac{d^2 y}{dx^2} = f(x, y, y') \tag{3.14}$$

and therefore we need two initial values, $y(x_0) = y_0$ and $y'(x_0) = y'_0$

Once again, we can start with a Taylor expansion of $y(x)$ (eqn (3.4)). The derivative $y'(x)$ can also be expressed as a Taylor expansion

$$y'(x + h) = y'(x) + hy''(x) + \frac{h^2}{2!} y'''(x) + \dots. \tag{3.15}$$

If terms containing $y'''(x)$ and above are ignored (due to the effect of h^2 and higher powers) in a first approximation, the expansions reduce to:

$$y(x + h) \approx y(x) + hy'(x) + \frac{h^2}{2!} y''(x)$$
$$y'(x + h) \approx y'(x) + hy''(x). \tag{3.16}$$

Since for this class of problem we know the initial values $y(x_0) = y_0$ and $y'(x_0) = y'_0$, we can calculate $y''(x_0)$.

The values $y(x_1)$ and $y'(x_1)$ can now be estimated from eqn (3.16):

$$y(x_1) = y(x_0) + hy'(x_0) + \frac{h^2}{2!} y''(x_0)$$
$$y'(x_1) = y'(x_0) + hy''(x_0).$$

Finally $y''(x_1)$ is calculated and the iteration process is repeated as required.

This is almost the second-order equivalent of the Euler–Cauchy method and the errors that are inherent could be determined by a series of small experiments which are left to the reader. Similarly, it should be possible to develop a predictor–corrector approach. We will however present an outline of the Runge–Kutta method as adapted for second-order initial-value problems.

The Runge–Kutta–Nyström method

This is a generalization of the Runge–Kutta method and can be applied to second order-differential equations. The steps in the algorithm will be outlined alongside a solution of the initial-value problem

$$\frac{d^2y}{dx^2} - \frac{dy}{dx} - x + y = 0 \text{ where } y \text{ and } y' \text{ are zero at } x = 0.$$

This differential equation can be expressed as

$$\frac{d^2y}{dx^2} = f(x, y, y') = x - y + \frac{dy}{dx}$$

The Runge–Kutta–Nyström algorithm uses an additional two auxiliary quantities, K and L and proceeds as detailed in Table 3.2.

Values of y_{n+1} can be determined by means of starting off with the values of $y(x_0)$, $y'(x_0)$ and proceeding piecewise thereafter up to the required value of n.

Table 3.2 The Runge–Kutta–Nystrom procedure

Algorithm	Example Solution of $f(x, y, y') = x - y + y'$
$x_{n+1} = x_n + h$	$x_{n+1} = x_n + 0.2$ (where $h = 0.2$)
$k_1 = \frac{h}{2}f(x_n, y_n, y'_n)$	$k_1 = 0.1(x_n - y_n + y'_n)$
define $K = \frac{h}{2}(y'_n + k_1/2)$	$K = 0.1(y'_n + 0.05(x_n - y_n + y'_n))$
$k_2 = \frac{h}{2}f[(x_n + h/2), (y_n + K), (y'_n + k_1)]$	$k_2 = 0.1[(x_n + 0.1) - (y_n + K) + (y'_n + k_1)]$ which can be expanded by inserting the appropriate values of K and k_1
$k_3 = \frac{h}{2}f[(x_n + h/2), (y_n + K), (y'_n + k_2)]$	$k_3 = 0.1[(x_n + 0.1) - (y_n + K) + (y'_n + k_2)]$ which can be expanded by inserting the appropriate values of K and k_2
define $L = h(y'_n + k_3)$	$L = 0.2(y'_n + k_3)$
$k_4 = \frac{h}{2}f[(x_n + h), (y_n + L), (y'_n + 2k_3)]$	$k_4 = 0.1f[(x_n + h) - (y_n + L) + (y'_n + 2k_3)]$ which can be expanded by inserting the appropriate values of L and k_3.
$y_{n+1} = y_n + h\left\{y'_n + \frac{1}{3}[k_1 + k_2 + k_3]\right\}$	
$y'_{n+1} = y'_n + \frac{1}{3}[k_1 + 2k_2 + 2k_3 + k_4]$ which is also used as an auxiliary for the next iteration.	

3.3 Techniques for partial differential equations

There are many situations in nature where a property is a function of more than one variable. One well-known example is the behaviour of a gas inside a balloon which is summarized by the Boyle–Charles law, $PV = RT$. The volume of the gas decreases with increasing pressure and expands with increasing temperature. Thus one could write that $V = F(P, T)$. The question of derivatives is simplified by allowing only one of the variables to change while the other is kept constant. For the case of a gas these are $\left(\frac{\partial V}{\partial T}\right)_P$ where the pressure is held constant and $\left(\frac{\partial V}{\partial P}\right)_T$ where the temperature is held constant.

The subject of thermodynamics is rich with functions of many variables and their derivatives. One can take the first law of thermodynamics, which states that '*the energy of the universe is constant*'. This means that the total energy, E of a system which has done some work can be quantified. One simple example is the state of a gas inside a cylinder (fitted with a frictionless piston) which has absorbed a quantity of heat ΔH and which has then expanded against the external atmospheric pressure. This can be summarized as

$$E = \Delta H + P\Delta V. \tag{3.17}$$

One might then ask how this might vary as a function of temperature. The total energy is in fact $E(P, V)$ so that one can differentiate it with respect to either variable. This leads to the following useful definitions:

$$C_P = \left(\frac{\partial E}{\partial T}\right)_P \quad \text{the specific heat at constant pressure;}$$
$$C_V = \left(\frac{\partial E}{\partial T}\right)_V \quad \text{the specific heat at constant volume.} \tag{3.18}$$

These are related through the gas constant by $C_P - C_V = R$ (the gas constant).

The ratio $\gamma = C_P/C_V$ is an important property in the adiabatic expansion of gases and gives an indication of the ease with which a gas can be liquefied.

The previous section indicated how partial derivatives appear in some of the apparently simplest phenomena. If partial derivatives, then why not partial differential equations? There are in fact a wide variety of partial differential equations and following an attempt to classify them the succeeding sections of this chapter will outline ways in which they may be solved by finite-difference methods.

A general second-order partial differential equation can be expressed as:

$$A\left(\frac{\partial^2 u}{\partial x^2}\right) + 2B\left(\frac{\partial^2 u}{\partial x \partial y}\right) + C\left(\frac{\partial^2 u}{\partial y^2}\right) = F\left[x, y, u, \frac{\partial u}{\partial x}, \frac{\partial u}{\partial y}\right] \tag{3.19}$$

where A, B, and C are coefficients (as in a quadratic equation) which may be constants or functions of x and/or y. Three different classes of differential equation can then be identified:

(1) $\qquad AC - B^2 > 0 \qquad\qquad$ *elliptic* equations;

(2) $\qquad AC - B^2 = 0 \qquad\qquad$ *parabolic* equations;

(3) $\qquad AC - B^2 < 0 \qquad\qquad$ *hyperbolic* equations.

The solution of any of these classes of differential equations may involve initial conditions. There will almost certainly be some boundary conditions in many of the problems which could be modelled. This leads to the further classification of problem types into Dirichlet, Neumann, and mixed as discussed in Section 2.2.1.

3.3.1 Finite-difference solutions of elliptic differential equations

This type of differential equation include two very important examples. Laplace's equation can be used to express the potential distribution in a charge-free region with boundaries at fixed potentials. It can also be used to give the temperature distribution in steady-state thermal problems (see eqn (2.6)). Poisson's equation (given in one-dimension in eqn (2.25))

$$\frac{\partial^2 u}{\partial x^2} + \frac{\partial^2 u}{\partial y^2} = \rho, \tag{3.20}$$

where ρ is a constant, can be used to obtain the potential within a charge-filled region and is particularly important in the simulation of carrier transport in semiconductor devices.

Just as in eqn (3.4) one can write a Taylor expansion, except that one of the variables is kept constant

$$u(x + h, y) = u(x, y) + h\frac{\partial u}{\partial x} + \frac{h^2}{2!}\frac{\partial^2 u}{\partial x^2} + \frac{h^3}{3!}\frac{\partial^3 u}{\partial x^3} + \dots. \tag{3.21}$$

Similarly we can write

$$u(x - h, y) = u(x, y) - h\frac{\partial u}{\partial x} + \frac{h^2}{2!}\frac{\partial^2 u}{\partial x^2} - \frac{h^3}{3!}\frac{\partial^3 u}{\partial x^3} + \dots. \tag{3.22}$$

If these expressions are subtracted and if we neglect terms in h^3 and higher then we get:

$$\frac{\partial u}{\partial x} = \frac{1}{2h}[u(x + h, y) - u(x - h, y)]. \tag{3.23}$$

We can similarly derive

$$\frac{\partial u}{\partial y} = \frac{1}{2k}[u(x, y + k) - u(x, y - k)] \tag{3.24}$$

where k is the discretization in the y direction.

A similar sequence of steps can be used to obtain

$$\frac{\partial^2 u}{\partial x^2} = \frac{1}{h^2}[u(x + h, y) - 2u(x, y) + u(x - h, y)]$$

$$\frac{\partial^2 u}{\partial y^2} = \frac{1}{k^2}[u(x, y + k) - 2u(x, y) + u(x, y - k)]. \tag{3.25}$$

For the general expression of an elliptic equation we can also deduce that:

$$\frac{\partial^2 u}{\partial x \partial y} = \frac{1}{4hk}[u(x + h, y + k) - u(x - h, y + k); -u(x + h, y - k)$$

$$+ u(x - h, y - k)]. \tag{3.26}$$

The finite-difference solution of the Laplace equation can be obtained by the summation of the two terms in eqn (3.25) and equating them to zero:

$$\frac{1}{h^2}[u(x + h, y) - 2u(x, y) + u(x - h, y)] + \frac{1}{k^2}[u(x, y + k) - 2u(x, y)$$

$$+ u(x, y - k)] = 0. \tag{3.27}$$

This corresponds to a situation where the discretizations in the x and y directions are not equal. If $h = k$ then eqn (3.27) simplifies considerably:

$$u(x + h, y) + u(x - h, y) + u(x, y + h) + u(x, y - h) = 4u(x, y) \tag{3.27}$$

or

$$u(x, y) = \frac{1}{4}[u(x + h, y) + u(x - h, y) + u(x, y + h) + u(x, y - h)]. \tag{3.28}$$

This is an extraordinary result and indicates that the value at any point is in fact the average of only the four points around it as is shown in Fig. 3.1.

Methods for solving the Laplace equation
(a) Simple relaxation

We can start by considering a typical example where a charge-free region of space is surrounded by two electrodes (see Fig. 3.2).

The definition of $V(x, y)$ along the boundary satisfies the criterion for a Dirichlet problem. The Laplace equation will hold for all mesh points but at

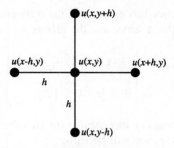

Fig. 3.1 A finite-difference representation of a region in space which is involved in the 5-point approximation of eqn (3.28).

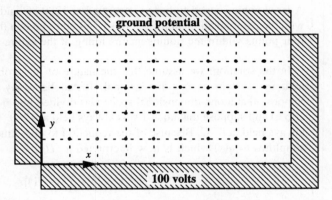

Fig. 3.2 A series of equi-distant mesh points in a charge-free region surrounded by two electrodes. The function $u(x,y)$ is then interpreted as $V(x,y)$.

the start of the solution these are unknown. Using a unity discretisation ($h = 1$) one could start with an 8×4 array, $_kA$ and perform the sequence of operations:

$$_kB = D \, _kA \qquad (3.29)$$

where D is the differential operation defined by eqn (3.28);

$$_{k+1}A = {_k}B. \qquad (3.30)$$

The subscripts k and $k + 1$ denote successive timesteps.

These two steps could be repeated until there is negligible difference between the value of the elements of $_kA$ and $_{k+1}A$. When this situation is achieved, every point will be the average of the four surrounding points so that eqn (3.28) would be fulfilled.

In practice this means that the differential operation is carried out for every point in the A array and the results are stored in the B array. Thus:

$$_1V(1, 1) = \frac{1}{4}[V(1 + 1, 1) + V(1 - 1, 1) + V(1, 1 + 1) + V(1, 1 - 1)]$$

$$= \frac{1}{4}[0 + 0 + 0 + 100] = 25 \text{ V}$$

This process can be repeated for all values of x and y for many iterations, but the convergence will be slow.

Example 3.2 A one-dimensional example of simple relaxation
A large cubic shaped furnace is operated at 1280 °C and is in equilibrium with the external ambient (0 °C). It is required to predict the temperatures at four points within the homogeneous lining at the centre of one face.

In the solution, we assume that the nature of the problem allows us to treat the system as one-dimensional. It can be easily solved by hand. Suppose that a one-dimensional space was divided into 6 nodes (numbered 1 to 6) and suppose that node 1 was always held at 1280 °C and node 6 was always held at 0 °C. Because of the reduction to one-dimension eqn (3.28) simplifies to $u(x)$ which is now interpreted as $T(x)$:

$$T(x) = \frac{1}{2}[T(x + 1) + T(x - 1)]. \tag{3.31}$$

The process of taking each point as the average of the two surrounding points according to eqns (3.30) and (3.31) is shown in Table 3.3.

Table 3.3 Simple relaxation procedure for a one-dimensional heat example

	furnace			lining		ambient	
	1280	0	0	0	0	0	$_1A$
$_1B$	1280	640	0	0	0	0	$_2A$
$_2B$	1280	640	320	0	0	0	$_3A$
$_3B$	1280	800	320	160	0	0	$_4A$
$_4B$	1280	800	480	160	80	0	$_5A$
$_5B$	1280	880	480	280	80	0	$_6A$
$_6B$	1280	880	580	280	140	0	$_7A$
$_7B$	1280	and	so	on	...	0	$_8A$

Convergence is much faster if one uses the single-step process

$$_{k+1}A = D_kA. \tag{3.32}$$

This means that the results of a calculation are immediately used to calculate the value at the next mesh-point. With the previous method $_1V(2,1)$ of Figure 3.2 would also be 25 volts. When this approach is used the expression then becomes.

$$_1V(2, 1) = \frac{1}{4}[V(2 + 1, 1) + V(2 - 1, 1) + V(2, 1 + 1) + V(2, 1 - 1)]$$

$$= \frac{1}{4}[V(2 + 1, 1) + {}_1V(1, 1) + V(2, 1 + 1) + V(2, 1 - 1)]$$

$$= \frac{1}{4}[0 + 25 + 0 + 100] = 31.25 \text{ V}.$$

The value that was calculated at the previous instant has been shown in bold.

One might feel that the low-value mesh points (near $x = 1$, $y = 1$) have the greater accuracy and the high-value mesh points will take longer to converge. This may be so, but if it is felt important then the direction of calculation can be reversed at every alternative iteration (i.e. from $x = 1$ to 8, $y = 1$ to 4 every odd iteration and from $x = 8$ to 1 and $y = 4$ to 1 every even iteration).

Example 3.3 A one-dimensional example with a faster relaxation method The problem in Example 3.2 can be revisited using this technique and a comparison of the results confirms the much faster progression towards convergence.

Table 3.4 A faster relaxation procedure for the heat example

	furnace		lining			ambient
$_1A$	1280	0	0	0	0	0
$_2A$	1280	640	320	160	80	0
$_3A$	1280	800	480	280	140	0
$_4A$	1280	880	580	360	180	0
$_5A$	1280	and	so	on	...	0

(b) Successive over-relaxation

The rate of convergence can be increased very significantly by means of a technique that involves the use of a residual matrix which is given by:

$$_kR = {}_kA - {}_{k-1}A. \tag{3.33}$$

The objective is to use a multiple of the residual when calculating the values at the next iteration

$$_{k+1}A = D_kA + B_kR. \tag{3.34}$$

If $B = 0$ then the situation is as before (simple relaxation). If on the other hand B is close to unity then the results can oscillate and if it is increased significantly, the system could become unstable. The objective is to choose a value of B (typically in the range $0 < B < 1$) which yields a critical damping response, i.e. the most rapid convergence. The value B_{optimal} is found to depend on the discretization dimensions of the problem. Although the value of B can be predicted analytically for certain problems there is no universally applicable formula. Thus a trial-and-error approach is recommended.

Some of these ideas can be demonstrated by applying eqns (3.32)–(3.34) to the problem outlined in Fig. 3.2. Tables 3.5–3.7 list the potentials $V(1, 1)$–$V(4, 1)$ (with respect to $x = 0$, $y = 0$ in the bottom left-hand corner) for the first five iterations for different values of B.

Table 3.5 Over-relaxation with $B = 0.0$

k	V(1,1)	V(2,1)	V(3,1)	V(4,1)
1	25.00	31.25	32.81	33.20
2	34.37	44.14	46.97	47.80
3	38.86	50.75	54.54	55.77
4	41.41	54.71	59.21	60.80
5	43.04	57.32	62.39	64.37

Table 3.6 Over-relaxation with $B = 1.0$

k	V(1,1)	V(2,1)	V(3,1)	V(4,1)
1	25.00	31.25	32.81	33.20
2	59.3	128.88	147.99	155.67
3	59.30	128.88	147.99	155.67
4	96.3	155.89	189.8	206.99
5	94.42	160.48	208.14	238.55

This continues to diverge to ever larger values

Table 3.7 Over-relaxation with $B = 0.5$

k	V(1,1)	V(2,1)	V(3,1)	V(4,1)
1	25	31.25	32.81	33.2
2	47.87	62.89	68.06	69.67
3	56.05	78.88	87.98	91.46
4	56.35	80.85	92.30	97.57
5	53.93	77.14	89.04	95.55

At this point, since we are discussing computable solutions it is relevant to include another hard-code example. The 'True Basic' programme which is included below could be used to monitor the convergence of $V(3,1)$ for different values of B. It could also be used to demonstrate that for this particular geometry the point in question converges fastest with $B = 0.15$. It should be noted that since the True Basic language cannot handle zero values of x and y, one has been added to each index so that the required results from the programme appear as $V(4,1)$.

* *

```
REM 2-DIMENSIONAL LAPLACE EQUATION SOLUTION BY SOR
DIM VNEW(10,6), VOLD(10,6), ERR(10,6)


REM  - - - - - - data initialisation - - - - - -
LET B = 0.2
   FOR I = 2 TO 9
     LET VNEW(I,1) = 100
     LET VNEW(I,6) = 0
   NEXT I

   FOR J = 2 TO 5
     LET VNEW(1,J) = 0
     LET VNEW(10,J) = 100
   NEXT J
REM - - - - - - end of data initialisation - - - - - -


REM  - - - - - - data transformation - - - - - -
   FOR K = 1 TO 500
     CALL fd
     CALL swap
   NEXT K
REM - - - - - - end of data transformation - - - - - -


REM  - - - - - - data output - - - - - -
   CALL printout
REM - - - - - - end of data output - - - - - -


REM  - - - - - - subroutines  - - - - - -

SUB fd
   FOR J = 2 TO 5
     FOR I = 2 TO 9
       LET VNEW(I,J)  =0.25*(VNEW(I-1,J)
                          +VNEW(I+1,J)+VNEW(I,J-1)
                          +VNEW(I,J+1))+B*ERR(I,J)
```

```
              LET ERR(I,J) = VNEW(I,J) - VOLD(I,J)
         NEXT I
      NEXT J
      END SUB

      SUB swap
         FOR J = 2 TO 5
            FOR I = 2 TO 9
               LET VOLD(I,J) = VNEW(I,J)
            NEXT I
         NEXT J
      END SUB

      SUB printout
         FOR J = 2 TO 5
            FOR I = 2 TO 9
               PRINT VNEW(I,J)
            NEXT I
         NEXT J
      END SUB
      REM  - - - - - - end of subroutines - - - - - -

      END
```
* *

In the code above it will be noted that the time index k was allowed to run to a maximum of 500. This was far in excess of what was needed for most values of B. Convergence at position (x, y) can be defined as the point in the iterative process when the residual, $_{k+1}V(x, y) - {_k}V(x, y)$ becomes less than some prescribed value. Depending on the required level of accuracy this could be 1%, 0.1%, or even less. We might reasonably define an overall convergence as the point when the residuals at all mesh points are less than the accuracy threshold.

These algorithms represent a set of linear equations and can easily be put into matrix form. They are classified as sparse matrices in that they have only a limited number of non-zero elements. This somewhat limits the techniques which are available for solution using matrix methods. The first node point, $V(1,1)$ in Fig. (3.2) can be expressed using eqn (3.28) as:

$$4\,_{k+1}V(1, 1) = {_{k+1}}V(0, 1) + {_k}V(2, 1) + {_k}V(1, 2) + {_{k+1}}V(1, 0)$$
$$= 0 + {_k}V(2, 1) + {_k}V(1, 2) + 100$$

or

$$_{k+1}V(1, 1) = 0.25{_k}V(2, 1) + 0.25{_k}V(1, 2) + 25$$

which is one equation in a Gauss–Seidel iteration (see Appendix), which is indeed what the algorithms above represent.

Explicit vs. implicit methods

Equations (3.30) and (3.31) represented an *explicit* method because the new value at each mesh point was calculated only in terms of the previously obtained values of the surrounding nodes. Equations (3.32)–(3.34) are *semi-implicit* methods because the new value at each point $_{k+1}V(x, y)$ use up to two values which have been calculated during the same iteration. It is also possible to calculate $_{k+1}V(x, y)$ using surrounding values at the same iteration, which have been simultaneously calculated. This would be termed an *implicit* method.

The ADI method

The Alternate Direction Implicit (ADI) method involves the arrangement of data in a special way so as to obtain a tri-diagonal matrix:

$$_{k+1}V(x - 1, y) - 4_{k+1}V(x, y) + {}_{k+1}V(x + 1, y)$$
$$= - {}_kV(x, y - 1) - {}_kV(x, y + 1) \qquad (3.35)$$

This contains only $k + 1$ terms on the left and if y is held constant and x is allowed to vary this represents a tri-diagonal matrix, which is shown in part below for $y = 2$.

$$\begin{bmatrix} 0 & -4V(1,2) & V(2,2) & - & -- & - & - & - & - & - \\ -- & V(1,2) & -4V(2,2) & V(3,2) & - & - & - & - & - & - \\ -- & - & V(2,2) & -4V(3,2) & V(4,2) & - & - & - & - & - \\ -- & - & - & V(3,2) & -4V(4,2) & V(5,2) & - & - & - & - \\ -- & - & - & - & V(4,2) & -4V(5,2) & V(6,2) & - & - & - \\ -- & - & - & - & - & V(5,2) & -4V(6,2) & V(7,2) & - & - \\ -- & - & - & - & - & - & V(6,2) & -4V(7,2) & V(8,2) & - \\ -- & - & - & - & - & - & - & V(7,2) & -4V(8,2) & 100 \end{bmatrix}$$

The numbers at the two extremes of the diagonal are of course the defined conditions on the boundaries.

So eqn (3.35) can be solved by Gaussian elimination and this can be repeated for every value of y.

At the next step the data can be arranged in yet another way:

$$_{k+2}V(x, y - 1) - 4_{k+2}V(x, y) + {}_{k+2}V(x, y + 1)$$
$$= - {}_{k+1}V(x - 1, y) - {}_{k+1}V(x + 1, y) \qquad (3.36)$$

This time the value of a set of tri-diagonal matrices are solved for a series of fixed values of x. The process of using eqns (3.35) and (3.36) at alternate iterations is repeated until the results converge to the required level of accuracy.

Just as the accelerating factor B was used in eqn (3.34) an acceleration factor, ω can be introduced into the ADI equations:

$$
\begin{aligned}
&_{k+1}V(x-1,y) - (2+\omega)_{k+1}V(x,y) + {}_{k+1}V(x+1,y) \\
&= -{}_kV(x,y-1) + (2-\omega)_kV(x,y) - {}_kV(x,y+1) \\
&_{k+2}V(x,y-1) - (2+\omega)_{k+2}V(x,y) + {}_{k+2}V(x,y+1) \\
&= -{}_{k+1}V(x-1,y) + (2-\omega)_{k+1}V(x,y) - {}_{k+1}V(x+1,y)
\end{aligned}
\tag{3.37}
$$

The value of the accelerating factor is given by $\omega = 2\sin(\pi/M)$, where M is the larger of the maximum discretization in the x and y direction, x_{max+1} or y_{max+1}. It is also possible to increase ω during the iterative process.

Solution of Poisson's equation

The techniques which have been outlined above can also be applied to the solution of Poisson's equation (3.20). The general form of the finite-difference approximations for equal x and y discretizations is:

$$
u(x,y) = \frac{1}{4}[u(x+h,y) + u(x-h,y) + u(x,y+h) + u(x,y-h)] - \frac{h^2}{4}\rho.
\tag{3.38}
$$

Further examples involving the solution of Poisson's equation by other techniques will be considered is later chapters.

The treatment of uneven boundaries

Figure 2.12 covered some examples where the level of complexity could render a problem intractable to analytical treatments. We saw that numerical solutions of Laplace's equation for these cases were possible, but the nature of the boundaries nevertheless presented a problem. The outer boundary in Fig. 2.12(a) can be approximated as shown in Fig. 3.3(a). A dense mesh is required in order to make any accurate approximation The alternative is to use mesh lines of unequal length as shown in Fig. 3.3(b). The most general case is illustrated in Fig. 3.4. In this case, which is similar to Fig. 3.1 except that the mesh point is surrounded by four points, P_1, P_2, P_3 and P_4 whose distances from the centre are ah, bh, ch, and dh, where h is the basic discretization. An analysis similar to that given in eqns (3.21)–(3.25) yields:

$$
\nabla^2 V(P_0) = \frac{2}{h^2}\left[\frac{V(P_1)}{a(a+c)} + \frac{V(P_2)}{b(b+d)} + \frac{V(P_3)}{c(c+a)} + \frac{V(P_4)}{d(d+b)} - \frac{(ac+bd)V(P_0)}{abcd}\right].
\tag{3.39}
$$

(a)   (b)

Fig. 3.3 (a) Cartesian approximation to a curve. (b) Improvement of the approximation by the use of unequal mesh lines (shown dashed).

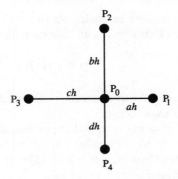

Fig. 3.4 Five-point finite-difference representation with unequal distances ($a h$ is a fraction of some discretization h).

For the Laplace equation one then obtains

$$V(P_0) = \frac{abcd}{(ac + bd)}\left[\frac{V(P_1)}{a(a + c)} + \frac{V(P_2)}{b(b + d)} + \frac{V(P_3)}{c(c + a)} + \frac{V(P_4)}{d(d + b)}\right]. \quad (3.40)$$

In a problem such as that outlined in Fig. 2.12(b) there is a need to store the value of a, b, c, and d for every node point. The modeller must decide whether this additional effort provides a worthwhile improvement on the approximation in Fig. 3.3 (a).

3.3.2 Finite difference solutions of parabolic differential equations

The two previous sections dealt with a very important class of partial differential equations. In many modelling applications they could be considered to represent the situation at equilibrium. It may be, however, that the current state has not existed for all previous time and we know intuitively that natural phenomena rarely undergo discontinuous transitions. It could be said that much modelling is concerned with the investigation of transient states

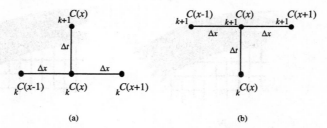

Fig. 3.5 Four-point (a) backward and (b) forward difference representations.

and in this context parabolic equations play a very significant part. Fick's second law of diffusion is an important example of this type of equations.

$$D \nabla^2 C(x, t) = \frac{\partial C(x, t)}{\partial t} \qquad (3.41)$$

where $C(x, t)$ is the instantaneous concentration of diffusant at position x and D is the diffusion constant.

Much of what has been said in previous sections could be applied, but there are stability problems and these must be considered in some detail here. For the purposes of presentation eqn (3.41) will be treated in one dimension with a unity value for D:

$$\frac{\partial^2 C(x, t)}{\partial x^2} = \frac{\partial C(x, t)}{\partial t}. \qquad (3.42)$$

This can be presented in finite difference form where the space and time discretizations are Δx and Δt respectively.

$$\frac{1}{\Delta t}[_{k+1}C(x) - {}_k C(x)] = \frac{1}{\Delta x^2}[_k C(x + 1) - 2_k C(x) + {}_k C(x - 1)] \qquad (3.43)$$

or

$$_{k+1}C(x) = {}_k C(x) + \frac{\Delta t}{\Delta x^2}[_k C(x + 1) - 2_k C(x) + {}_k C(x - 1)]. \qquad (3.44)$$

This is in fact a four-point backward difference scheme and it is shown diagramatically in Fig. 3.5. It is explicit in that the value for $_{k+1}C(x)$ is based only on values which are obtained at the previous iteration. An explicit scheme such as this will not be stable unless $\Delta t/\Delta x^2 \leq 1/2$ which can cause a significant computational load in a simulation. Thus, if Δx is 0.1 then Δt must be less than 0.005 and if the spatial discretization is reduced by a factor 2 then the number of iterations for a given time interval must increase fourfold. Further, when $D \neq 1$ then this stability criterion reads as:

$$D\frac{\Delta t}{\Delta x^2} \le \frac{1}{2}.$$

It is also interesting to observe that when $D\frac{\Delta t}{\Delta x^2} = \frac{1}{2}$ then the concentration at a point is the average of the value of the surrounding two points:

$$_{k+1}C(x) = \frac{1}{2}[_k C(x+1) + _k C(x-1)]. \tag{3.45}$$

The method of Liebmann uses a forward (implicit) difference scheme as outlined in Fig. 3.5(b) In this case $_{k+1}C(x)$ is calculated using $_k C(x)$, $_{k+1}C(x)$ and surrounding values at the new time step:

$$\frac{1}{\Delta t}[_{k+1}C(x) - _k C(x)] = \frac{1}{\Delta x^2}[_{k+1}C(x+1) - 2_{k+1}C(x) + _{k+1}C(x-1)] \tag{3.45}$$

or

$$_{k+1}C(x) = _k C(x) + \frac{\Delta t}{\Delta x^2}[_{k+1}C(x+1) - 2_{k+1}C(x) + _{k+1}C(x-1)]. \tag{3.46}$$

In this case there are no instability problems although one has to solve a set of linear equations at every iteration.

There are several other schemes which succeed in avoiding stability problems and these are considered below.

Crank–Nicholson method

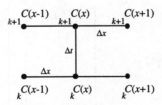

Fig. 3.6 Six-point Crank–Nicholson difference scheme.

This is an implicit technique which uses a six-point difference (see Fig. 3.6). The spatial second derivative is written in terms of $_k C(x)$ and $_{k+1}C(x)$. For convenience the ratio $D\frac{\Delta t}{\Delta x^2}$ is replaced by r, so that

$$_{k+1}C(x) - _k C(x) = \frac{r}{2}[_k C(x+1) - 2_k C(x) + _k C(x-1)]$$
$$+ \frac{r}{2}[_{k+1}C(x+1) - 2_{k+1}C(x) + _{k+1}C(x-1)] \tag{3.47}$$

or

$$(2 + 2r)_{k+1}C(x) - [_{k+1}rC(x+1) + _{k+1}C(x-1)]$$
$$= (2 - 2r)_kC(x) + r[_kC(x+1) + _kC(x-1)]. \quad (3.48)$$

The terms on the right-hand side of eqn (3.48) are known. If there are N discretization points in the model then there are $N - 1$ linear equations for the $N - 1$ unknowns.

Although there is no stability restriction on the value or r, we must recognize that the larger the value of r, the larger is the possible deviation from the true results. The situation where $r = 1$ is of particular interest as it yields a result similar to eqn (3.28):

$$_{k+1}C(x) = \frac{1}{4}[_kC(x+1) + _kC(x-1) + _{k+1}C(x+1) + _{k+1}C(x-1)]$$

(3.49)

Every point $_{k+1}C(x)$ is the average of the sum of surrounding points at the present and previous iterations.

The duFort–Frenkel method

This is a five-point difference scheme which includes the value, $_{k-1}C(x)$ from the previous-but-one iteration:

$$(1 + 2r)_{k+1}C(x) = 2r[_kC(x+1) + _kC(x-1)] + (1 - 2r)_{k-1}C(x) \quad (3.50)$$

or in matrix form

$$_{k+1}\mathbf{C} = \frac{2r}{1+2r}\mathbf{A}_k\mathbf{C} + \frac{1-2r}{1+2r}{}_{k-1}\mathbf{C} + {}_0\mathbf{C} \quad (3.51)$$

where
$$_{k+1}\mathbf{C} = \begin{array}{c} \\ \\ \\ \\ \\ k+1 \end{array}\begin{bmatrix} C(1) \\ - \\ - \\ - \\ - \\ C(N-1) \end{bmatrix} \qquad _0\mathbf{C} = \begin{bmatrix} C(0) \\ - \\ - \\ - \\ - \\ C(N) \end{bmatrix}$$

and
$$\mathbf{A} = \begin{bmatrix} 0 & 1 & \cdots & & \\ 1 & 0 & 1 & \cdots & \\ \vdots & 1 & 0 & \cdots & 1 \\ \vdots & & & & \\ & & \cdots & 1 & 0 \end{bmatrix}$$

This is explicit and unconditionally stable of all values of r, but there is a consistency criterion: it will only converge to the *exact* value if $\Delta t / \Delta x \to 0$ as $\Delta t, \Delta x \to 0$.

Example 3.4 Modelling the dispersion of a herbicide
This example which is due to Crank *et al.* [3.2] considers the dispersion of a herbicide which is applied to the top of a soil column to which water is added at a fixed rate.

The problem in fact draws on concepts which would be familiar to chemical engineers. At any point z in the column of depth L we can define a concentration $C(z)$ of chemical in the water. There will also be a concentration of chemical $S(z)$ which is attached (adsorbed) to the soil itself. These two quantities are related through the *adsorption isotherm K*:

$$S = KC.$$

We can also define:
$\theta =$ the volume of soil water per unit volume of soil;
$\rho =$ the dry density of the soil.
Thus, a unit volume of soil contains a quantity, q of chemical given by:

$$q = \theta C + \rho S = (\theta + \rho K)C.$$

The differential equation for the processes of dispersion (due to fluid flow) and diffusion is given by:

$$\frac{\partial C}{\partial t} = \bar{\Delta}\frac{\partial^2 C}{\partial z^2} - \bar{V}\frac{\partial C}{\partial z}$$

where $\bar{\Delta} = \dfrac{\Delta\theta}{\theta + \rho K'}$ (Δ is the dispersion coefficient) and $\bar{V} = \dfrac{\Delta v}{\theta + \rho K'}$,
(v is the velocity of water in the soil pores).

There are two Neumann boundary conditions which ensure that mass is conserved:

$$\frac{\partial C}{\partial z} = \frac{v}{\Delta}C \text{ at } z = 0$$

and

$$\frac{\partial C}{\partial z} = 0 \text{ at } z = L.$$

It is best to approach the solution to this problem using dimensionless variables. Thus we set:

$$T = \frac{\Delta t}{L^2}; \quad Z = \frac{z}{L}; \quad c = \frac{C(z,t)}{C_0}.$$

Setting $\alpha = \dfrac{Lv}{\Delta}$ one gets the differential equation in dimensionless form:

$$\frac{\partial^2 c}{\partial z^2} = \frac{\partial c}{\partial T} + \alpha \frac{\partial c}{\partial z}$$

with boundary conditions

$$\frac{\partial c}{\partial z} = \alpha c \text{ at } z = 0$$

and

$$\frac{\partial c}{\partial z} = 0 \text{ at } Z = 1.$$

Dimensionless space and time is discretized (in units of ΔZ from 1 to Z_{max}. ΔT is the dimensionless time. Derivatives are then replaced by their finite-difference approximations. Using $r = \Delta T / \Delta Z^2$ and $s = \alpha \Delta T / 2 \Delta Z$ we get:

$$_{k+1}c(Z) = (r+s)\,_kc(Z-1) + (1-2r)\,_kc(Z) + (r-s)\,_kc(Z+1).$$

In order to apply the boundary conditions, we must assume fictitious points, $_kc(-1)$ and $_kc(Z+1)$ which lie beyond the boundaries (see Section 4.3.5 in the next chapter). The concentration at $Z = 0$ is then given by:

$$_{k+1}c(0) = (r+s)_kc(-1) + (1-2r)_kc(0) + (r-s)_kc(1).$$

But the boundary condition at $Z = 0$ says that

$$\frac{_kc(1) - _k c(-1)}{2\Delta Z} = \alpha_k c(0)$$

so that we can set $_kc(-1) =_k c(1) - 2\alpha \Delta Z_k c(0)$.

Therefore $_{k+1}c(0) = [1 - 2r - 2\alpha \Delta Z(r+s)]_k c(0) + 2r\,_kc(1)$.

A similar argument for the other boundary leads to

$$_{k+1}c(0) = 2r_kc(Z_{max} - 1) + (1-2r)\,_kc(Z_{max}).$$

Thus all points in the mesh are now defined and the value of $c(Z, T)$ and therefore $C(z, t)$ can be determined explicitly.

3.3.3 Finite-difference solutions of hyperbolic differential equations

The equation for wave motion is one of the most important examples that falls under this heading:

$$\frac{\partial^2 \psi}{\partial t^2} = c^2 \frac{\partial^2 \psi}{\partial x^2} \text{ (where } c \text{ is the velocity)}$$

(where c is the velocity)
which may have boundary conditions, conditions of initial displacement or initial velocity.

Explicit scheme

Equation (3.52) can be expressed as an explicit five-point difference scheme:

$$\frac{1}{\Delta t^2} \left[{}_{k+1}\psi(i) - 2{}_k\psi(i) + {}_{k-1}\psi(i) \right] = \frac{c^2}{\Delta x^2} \left[{}_k\psi(i+1) - 2{}_k\psi(i) + {}_k\psi(i-1) \right].$$

(3.53)

Using the definition $r^2 = \frac{\Delta t^2 c^2}{\Delta x^2}$ it can be shown that if $0 < r \leq 1$ then the routine is stable. When $r = 1$ eqn (3.51) simplifies to

$$_{k+1}\psi(i) = {}_k\psi(i+1) + {}_k\psi(i-1) - {}_{k-1}\psi(i).$$ (3.54)

Unlike most of the hyperbolic formulations this routine needs to be started. Some of the ideas which were developed earlier can be used here. For simplicity we will set $c = 1$.

From the initial condition $\frac{\partial_0 \psi(x)}{\partial x} = g(x)$ we obtain:

$$\frac{1}{2\Delta x} \left[{}_1\psi(i) - {}_{-1}\psi(i) \right] = g(i).$$ (3.55)

Thus

$$_{-1}\psi(i) = {}_1\psi(i) - 2\Delta x\, g(i).$$ (3.56)

This can be substituted into eqn (3.54) for time $k = 0$ which after rearrangement yields:

$$_1\psi(i) = \frac{1}{2} \left[{}_0\psi(i-1) - {}_0\psi(i+1) \right] + 2\Delta x\, g(i)$$ (3.57)

which now expresses $_1\psi(i)$ in terms of the initial conditions.

Implicit scheme

There are many different alternatives, but only one implicit scheme will be presented here. The time component is unchanged, but on this occasion the spatial derivatives are taken over three points in time, $k + 1$, k and $k - 1$.

$$_{k+1}\psi(i) - 2_k\psi(i) + _{k-1}\psi(i) = r^2\left[\frac{_{k+1}\delta^2_x}{4} + \frac{_k\delta^2_x}{2} + \frac{_{k-1}\delta^2_x}{4}\right] \quad (3.58)$$

where $_{k+1}\delta^2_x = _{k+1}\psi(i+1) - 2_{k+1}\psi(i) + _{k+1}\psi(i-1)$ and similarly for $_k\delta^2_x$ and $_{k-1}\delta^2_x$.

This scheme, which gives a tri-diagonal system is stable for all $r > 0$.

Example 3.5 Ultrasonic wave propagation in two dimensions

Delsanto and colleagues [3.3] have devised a finite-difference type formulation which is ideal for use on massively parallel computers. The example which is outlined here applies these concepts to ultrasonic propagation in a homogeneous isotropic plate. Complete symmetry is assumed both in the plate and the source with respect to the z axis so that the problem can be treated as two-dimensional.

A general displacement vector w has components u and v in the x and y directions with ρ as density and λ and μ as Lamé constants [3.4]. One can write two equations:

$$\rho\frac{\partial^2 u}{\partial t^2} = (\lambda + 2\mu)\frac{\partial^2 u}{\partial y^2} + \mu\frac{\partial^2 u}{\partial x^2} + (\lambda + \mu)\frac{\partial^2 v}{\partial xy}$$

$$\rho\frac{\partial^2 v}{\partial t^2} = (\lambda + 2\mu)\frac{\partial^2 v}{\partial x^2} + \mu\frac{\partial^2 v}{\partial y^2} + (\lambda + \mu)\frac{\partial^2 u}{\partial xy}.$$

These equations can be written in finite difference forms as

$$\begin{aligned}
_{k+1}u(i,j) = {}& \alpha\left[_k u(i-1,j) + _k u(i+1,j)\right] \\
& + \beta\left[_k u(i,j-1) + _k u(i,j+1)\right] - 2\xi_k u(i,j) - _{k-1}u(i,j) \\
& + \gamma\left[_k v(i-1,j-1) + _k v(i-1,j+1) + _k v(i+1,j-1)\right. \\
& \left. + _k v(i+1,j+1)\right]
\end{aligned}$$

and

$$\begin{aligned}
_{k+1}v(i,j) = {}& \beta\left[_k v(i-1,j) + _k v(i+1,j)\right] \\
& + \alpha\left[_k v(i,j-1) + _k v(i,j+1)\right] - 2\xi_k v(i,j) - _{k-1}v(i,j) \\
& + \gamma\left[_k u(i-1,j-1) + _k u(i+1,j-1) + _k u(i+1,j-1)\right. \\
& \left. + _k u(i+1,j+1)\right]
\end{aligned}$$

where

$$\alpha = \left[\frac{V_L}{h/\Delta t}\right]^2; \quad \beta = \left[\frac{V_T}{h/\Delta t}\right]^2; \quad \gamma = \frac{\alpha + \beta}{4}; \quad \xi = \alpha + \beta - 1;$$

$$V_L = \sqrt{\frac{\alpha + 2\mu}{\rho}}; \quad V_T = \sqrt{\frac{\mu}{\rho}}.$$

The best choice for stability occurs when the space and time discretizations are related as $h/\Delta t = \sqrt{2}V_L$ [3.5]. These equations can then be solved, but we once again come back to the usual problem of computation time versus level of detail. It is possible to speed up the process by using several computer processors in parallel so that each does part of the task. Delsanto and colleagues [3.3] used a 'massively parallel computer' called a *connection machine* which was configured so that each node of the modelling problem was mapped to an individual processor. Some of their results are shown in Fig. 3.7. Further examples of this approach can be found in [3.6, 3.7, 3.8].

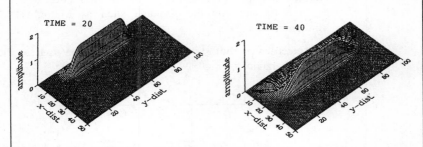

Fig. 3.7 The propagation of a longitudinal rectangular pulse as a function of time using a formulation based on Example 3.5 reprinted from [3.3] with permission.

References

3.1 F.D. Tabbutt, The Belousov–Zhabotinsky reaction, *Survey of Progress in Chemistry* **10** (1983) 129–187.

3.2 J. Crank, N.R. Mc Farlane, J.C. Newby, G.D. Paterson, and J.B. Pedley *Diffusion processes in environmental systems,* Macmillan 1981 pp. 105–113.

3.3 P.P. Delsanto, H.H. Chaskelis, R.B. Mignogna, T.V. Whitcombe and R.S. Schechter, Connection machine simulation of ultrasonic wave propagation: two-dimensional case, *Review of Progress in Quantitative Nondestructive Evaluation* **11** (1992) 113–120.

3.4 R. Feynman, *The Feynman lectures on physics* Vol II, Addison Wesley 1977.

3.5 W.H. Press, B.P. Flannery, S.A. Teukolsky and W.T. Vetterling, *Numerical recipes,* Cambridge University Press, 1986.

3.6 P.P. Delsanto, T. Whitcombe, H.H. Chaskelis and R.B. Mignogna, Connection machine simulation of ultrasonic wave propagation in materials. In the one-dimensional case, *Wave motion* **16** (1992) 65–80.

3.7 P.P. Delsanto, R.S. Schechter, H.H. Chaskelis, R.B. Mignogna and R. Klive, Connection machine simulation of ultrasonic wave propagation III the three dimensional case (to appear in *Wave Motion*).

3.8 R.S. Schechter, H.H. Chaskelis, R.B. Mignogna and P.P. Delsanto, Real time parallel computation and visualisations of ultrasonic processes in solids *Science* **265** (1994) 1188–1192.

Examples, exercises and projects

3.1 Use the Euler–Cauchy method to obtain a solution for $\frac{dy}{dx} = \frac{1}{x}$ in the range $0 \leq x \leq 4$, where the initial value, $y(0) = 1$. It is instructive to compare results for some value of x (e.g. $x = 2$) using two values of discretization (e.g. $h = 0.1$ and $h = 0.01$).

3.2 Develop a Runge–Kutta treatment for $\frac{dy}{dx} = x^3$, $y(0) = 0$ and attempt to express the resultant expression for y_{n+1} in a closed form. It should be apparent that the algebraic manipulation is even more cumbersome than the example quoted in the text. The derivation will be helped by the use of an symbolic package such as MathCAD, Maple, or Mathematica.

3.3 The electrical circuit shown in Fig. 3.8 can be used in the study of chaotic behaviour.

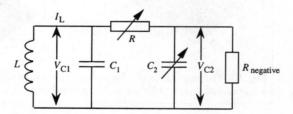

Fig. 3.8 Circuit for Exercise 3.3.

The governing equations are:

$$C_1 \frac{dV_{C1}}{dt} = R(V_{C1} - V_{C2}) - gV_{C1}$$

$$C_2 \frac{dV_{C2}}{dt} = R(V_{C1} - V_{C2}) + I_L$$

$$L \frac{dI}{dt} = -V_{C2}.$$

The conductance parameter g in the first of the equations above can be obtained from the graph Fig. 3.9.

Develop a polynomial expression for g and use it in a Runge–Kutta solution for V_{C1}, V_{C2}, and I in order to investigate the influence of R and C_2.

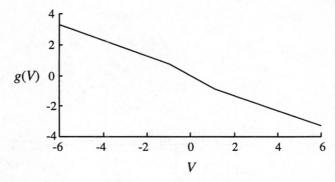

Fig. 3.9 Graph of g vs. V for Exercise 3.3.

3.4 The drawing shown in Fig. 3.10 is a schematic spark-plug ignitor for a gas disposal system. High voltage impulses are applied as shown.

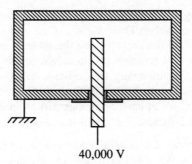

40,000 V

Fig. 3.10 Schematic drawing for a spark-plug igniter.

Make maximum use of symmetry to reduce the computational load for a relaxation solution of the Laplace equation. Arrange dimensions to ensure that the field at the tip exceeds 8×10^7 Vm^{-1}. The field at the entry point of the HT electrode is very high and an insulator must be used. This insulator has a breakdown of 10^8 Vm^{-1}. Calculate the required thickness of insulator in your design. If the insulator has a permittivity of 5 then one node in this material would be equivalent to 5 nodes in air. How would this change the design and if the insulator were extended into the enclosed space, would this lead to any improvements?

3.5 As a first approximation to a cooking process we could treat a thick cut of meat on a frying pan as a one dimensional diffusion of heat through a column of a water-like substance where the cellular structure inhibits convective heat transfer. The thermal diffusivity D of can be obtained

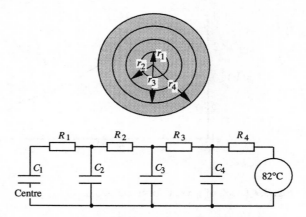

Fig. 3.11 Schematic drawing for a Christmas pudding being cooked and its analogue RC network.

using the density, specific heat and thermal conductivity (ρ, C_p, and k_T respectively): $D = k_T/\rho C_p$. One end of this model is connected to a constant temperature and the other is insulated. Investigate the cooking time as a function of the height of the joint above the pan and the pan temperature using two criteria: (i) The joint must not be considered cooked until the node furthest from the heat source has been above 95 °C for at least 5 minutes. (ii) No node must exceed 200 °C for more than 5 minutes or it will start to char.

3.6 Problem 2.8 considered a sphere of radius r radiating heat into free space. We will now proceed to see how this could be solved by applying finite-difference modelling to a simpler problem, namely the heating of a Christmas pudding.

For our purposes a Christmas pudding is a sphere of water where heat can transfer only by thermal conduction. It is divided into a set of 'onion rings' each of which has a thermal capacitance and a thermal resistance. Thus the concentric rings can be represented by a one-dimensional analogue RC network (see Fig. 3.12). If the width of an annulus is Δr then the volume of the innermost is $(4/3)\,\pi r_1^3$. The capacitance is then the product of density, the volume and the specific heat. The volume of the next annulus is $(4/3)\pi[(r_1 + \Delta r)^3 - r_1^3]$. The next one is $(4/3)\pi[(r_1 + 2\Delta r)^3 - (r_1 + \Delta r)^3]$ and so on. The resistance is given by $\Delta r/(k_T$ area) where k_T is the thermal conductivity. In calculating the area of any annulus there is a question of whether one should use the inner radius or the outer radius. It might be best to use the mean of the two or even the radius which bisects equal volumes within an annulus. These concepts should be developed into an algorithm to solve

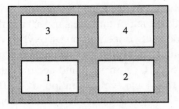

Fig. 3.12 An idealized dwelling with central heating.

a one-dimensional RC network which is open circuit at the inside and connected to a voltage source equivalent to $(100 \ ^\circ\text{C} - \text{room}$ temperature) at the other. One could then consider a pudding of 5 cm radius and where $\Delta r = 1$ cm. The cooking time could be estimated as the time which is required before the innermost capacitor has been at $99.9 \ ^\circ\text{C}$ for ten minutes. This result should then be compared with the cooking time for a pudding of 10 cm radius. Are there any conclusions that can be drawn that might be useful at Christmas time? As a further development you could examine the effects of pressure cooking where temperatures as high as $120 \ ^\circ\text{C}$ are possible.

3.7 Help restore domestic bliss by considering Fig. 3.12 which shows an idealized dwelling which has central heating.

Initially the entire structure is at the external ambient temperature $(10 \ ^\circ\text{C})$. For the purposes of the model it may be assumed that the volume of air in each room where a radiator is switched on represents a constant source which heats the walls, floor, and ceiling. Once the wall nodes next the surface have been within $1 \ ^\circ\text{C}$ of the radiator temperature for 20 minutes the radiator in a room switches off and the temperature at the surface of the wall becomes the source which keeps the room warm. If the requirement is to have a mean room temperature of $20 \ ^\circ\text{C}$ during a 24 hour period the arguments are based on how this should be achieved: should the water temperature in the radiator be warm or hot? These will lead to different thermal cycles and the objective is to determine which is the more economic in terms of heat input. In this context a 'warm' radiator will lead to a room temperature of $25 \ ^\circ\text{C}$ and a 'hot' radiator will lead to a room temperature of $30 \ ^\circ\text{C}$. The situation might be quite different depending on whether thermal or conventional bricks are used.

3.8 The high temperature oxidation of silicon is an interesting moving boundary problem. As the reaction progresses oxygen must diffuse through the silicon dioxide layer in order to reach the interface and undergo further reaction. This can be simulated by starting with a one-dimensional model representing the normal to the surface in a large area problem where edge effects can be ignored. The silicon dioxide is initially represented by 5 nodes and underlying silicon is represented by 50 nodes, each of unit area and 10 nm thickness (1 nm = 10^{-9} m). There is a constant concentration of oxygen on the outside of the oxide which constitutes a diffusion source. Oxygen diffuses through the oxide ($D = 6 \times 10^{-12}$ m^{2}s^{-1}) and accumulates in the node nearest the silicon until the local concentration is sufficient for a balanced reaction (Si + O$_2$ → SiO$_2$) to occur. When this happens the entire concentration of oxygen at this node is removed, the number of silicon nodes is reduced by one and the number of oxide nodes is increased by one. The diffusion/reaction process then continues and the objective is to model the growth of the oxide layer as a function of time. The simulation will require a knowledge of the number of silicon atoms within a node and this can be obtained from tables of density and atomic weight and the knowledge that 1(g) atomic weight of silicon represents 6.02×10^{23} atoms. It may also be assumed that the constant concentration of oxygen at the oxide surface is 1000 times this value.

3.9 A chemical reaction generates heat and this is a matter of concern to chemical engineers. In this problem it is required to simulate the temperature distribution within a flow reactor.

Reactants A and B are added in equal concentrations to a flow of water at one end of a deep reactor vessel consisting of two heavy metal plates set 2 cm apart. The initial concentration of each reactants is 5 moles (gram molecular weights) per 1000 cm^3. The reaction proceeds as:

$$A + B \longrightarrow C$$

Since the initial concentrations [A$_0$] and [B$_0$] are equal we can write the concentration, [C(t)] in terms of the rate constant, k as:

$$[C(t)] = [A_0]\left[1 - \frac{1}{1 + kt}\right]$$

If the reaction rate constant $k = 1$ and if the flow rate between the plates is 5 cm per second then the concentration of product produced at every point within the reactor will be constant and can be plotted.

The formation of product leads to the generation of heat (1000 J for every mole produced). This information can be used to estimate the amount of heat generated at any location within the reactor and since this will remain constant the thermal properties of water can be used in a two-dimensional Poisson equation to determine the temperature at every location assuming that the temperatures of the two plates remain unchanged.

How would you tackle the problem if the reaction rate k doubled in value for every 10 °C rise in temperature? What conditions might lead to thermal runaway and how could it be avoided?

Models based on distributed electromagnetic analogues

4

The previous chapter presented a series of methods for the solution of differential equations by means of finite difference approximations with the emphasis on techniques rather than applications. This is the conventional approach to modelling physical systems. In the present chapter, we consider an alternative approach which involves describing the physical system in terms of an electrical analogue. This circuit is then activated by the application of currents/voltages and the subsequent response provides a direct solution of the original physical problem. Before the advent of computers physical networks were constructed and appropriate measurements provided the required result [4.1].

Transmission Line Matrix (TLM) is a computerized implementation of this philosophy. It seeks to describe problems directly in terms of the time response of the telegraphers' equation which in general form can be described by:

$$\nabla^2 \phi = A \, \frac{\partial^2 \phi}{\partial t^2} + B \, \frac{\partial \phi}{\partial t}. \tag{4.1}$$

This is an embodiment of the fundamental laws of electromagnetics and describes the space and time behaviour of signals (digital or analogue) in a range of electrical networks. It is so called because it was first observed in relation to telegraph and telephone cables. Any physical problem which is observed to follow an equation of this form can be modelled by means of an equivalent electrical network. TLM is based firmly on electromagnetic principles, however many of its concepts may also be carried over into rule-based and probabilistic approaches, covered in Chapters 5 and 6 respectively.

The creation of a TLM model starts, as usual, with a discretization of the problem in space and/or time. Spatial discretization will again be a mirror of the physical model, but the approach to time discretization is somewhat different. A signal which enters a length of electrical transmission line takes a finite amount of time to reach the other end. Thus, if the problem is represented by a network with transmission lines then time discretization is enforced on the system since information cannot move from one part to another in zero time. If the velocity of propagation is constant throughout the model then the spatial separation between observation points define the unit of time discretization.

The telegraph equation is not easy to treat in a generalized way, and modellers tend to reserve TLM for extreme situations where one or other term of eqn (4.1) becomes insignificant.

Case 1: The first term dominates and the second is negligible. This is equivalent to a hyperbolic equation:

$$\nabla^2 \phi \approx A \frac{\partial^2 \phi}{\partial t^2}. \tag{4.2}$$

Case 2: The second term dominates and the first is negligible. This is in effect a parabolic equation:

$$\nabla^2 \phi \approx B \frac{\partial \phi}{\partial t}. \tag{4.3}$$

In this chapter we will outline the basics of TLM techniques and will demonstrate areas of application at these extremes. We will attempt to highlight both the strengths and the weaknesses which might guide the modeller towards or away from TLM as a modelling tool.

4.1 Transmission lines

A transmission line is an entity which has electrical capacitance, inductance and resistance distributed along its length[†]. The inductance, capacitance and resistance will depend on that length, so it is useful to use distributed properties, L_d, C_d, and R_d where $C_d = C_{total}/\text{Length}$, etc. It is often convenient to collect these distributed parameters over a unit length and display them as if they were lumped into a series of discrete components. Figure 4.1 shows a lumped parameter representation of a lossless ($R_d = 0$) transmission line. Figure 4.2 shows a small portions of a television coaxial cable and the down-feed for an FM radio receiver internal aerial, both of which can be successfully treated as lossless transmission lines so long as the overall length is not such as to severely attenuate the signal.

The distributed capacitance and inductance in a lossless transmission line can be used to deduce two important parameters, namely the impedance Z and the velocity v of an impulse on the line:

$$Z = \sqrt{\frac{L_d}{C_d}}$$
$$v = \sqrt{\frac{1}{L_d C_d}}. \tag{4.4}$$

[†] Strictly speaking, there will be time delay in any real electrical component, but for normal analysis devices such as resistors, capacitors, and inductors are treated as ideal so that the inherent delay is ignored and only the delay in transmission lines is considered.

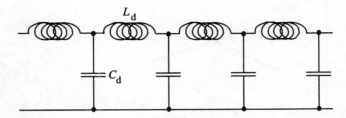

Fig. 4.1 Four sections of a lumped parameter representation of a lossless transmission line.

The time taken for an electrical signal impressed on the line to traverse it will be determined by the local velocity of light, which depends on the permittivity of the material which makes up the line.

An electrical impulse travelling on a network of lossy transmission lines will obey the telegraph equation for electromagnetic propagation; the generalization of eqn (4.1) for an RLC distributed network can be expressed as:

$$\nabla^2 \phi = \alpha \, L_\mathrm{d} C_\mathrm{d} \frac{\partial^2 \phi}{\partial t^2} + \beta \, R_\mathrm{d} C_\mathrm{d} \frac{\partial \phi}{\partial t} \tag{4.5}$$

where $\alpha = 1, 2, 3$ depending on dimension and $\beta = 2\alpha$. The first term on the right-hand side of this equation represents lossless propagation of a wave or impulse. The second term accounts for the effects of power dissipation in the distributed resistance. In the following sections the two extremes will be treated as limiting cases.

4.2 Lossless TLM models

Here we consider only the case where the first term in eqn (4.5) dominates. In electromagnetic terms this investigates the electric and magnetic field components (E_x, E_y, E_z, H_x, H_y, H_z) of the equations of James Clerk Maxwell.

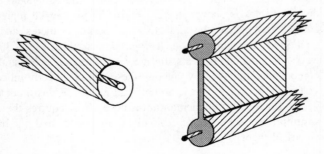

Fig. 4.2 Two examples of lossless transmission lines.

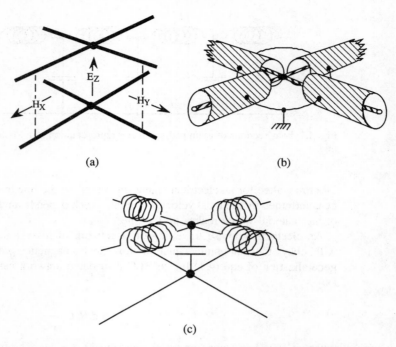

Fig. 4.3 (a) Intersecting pair of two-wire lossless transmission lines. (b) Practical example involving intersecting pair of co-axial cables. (c) Lumped circuit equivalent.

These are not easy to handle, but can be simplified by working in two dimensions when we can set many of the components equal to zero. There are two obvious cases: shunt mesh and series mesh representations.

4.2.1 Shunt mesh representation

Figure 4.3(a) shows a single two dimensional cell formed of transmission lines, where each arm links to other similar cells. The entire space of the problem is filled out as a square mesh of these cells. The equivalent arrangement using coaxial cables is shown in Fig. 4.3(b). These can be represented by lumped components where the inductors are series elements and the capacitor is connected across the lines as a *shunt* element as in Fig. 4.3(c). In terms of the Maxwell field components this network yields H_x, H_y, and E_z and all other components are zero. By analogy these can be directly related to currents and voltage in a circuit. This is an essential feature of this modelling approach.

The practical implementation of these ideas requires the use of two simple assumptions, one equation from electromagnetics and one theorem from basic electrical theory:

1. The first assumption is that all data (field amplitudes) are represented by impulses of very short duration. Thus an impulse entering what it perceives to be an infinite transmission line has no knowledge about the length of the line; indeed it is unaware that the line is finite until it arrives at a discontinuity.

2. The second assumption is needed for computational purposes and requires that all pulses move around the spatial mesh in synchronism. This is a feature of the solution technique rather than any physical aspect.

3. The electromagnetic equation is that for the reflection of a wave (impulse) at a discontinuity. If an impulse, travelling on a transmission line (of impedance Z) arrives at a discontinuity where the impedance becomes Z_T then the reflection coefficient is given by:

$$\rho = \frac{Z_T - Z}{Z_T + Z}. \qquad (4.6)$$

In the network shown in Fig. 4.3(b) the reflection coefficient, ρ is equal to −0.5 since a pulse arriving at the intersection from any arm will see a termination consisting of three 'infinitely' long transmission lines in parallel.

4. The final item that is required is Thévenin's theorem which describes the equivalent circuit of a transmission line in terms of the open-circuit voltage and the short-circuit impedance. The impedance is an easy matter: if we look in at one end of a transmission line with the other end shorted, then the measured impedance is Z. In order to visualise the open-circuit voltage, we must consider an incoming impulse of magnitude ^iV as it approaches the observation point. By the Thévenin definition this is an open circuit ($Z_T = \infty$) and thus $\rho = 1$. Therefore, at the instant that the impulse arrives, we have the superposition of the incident (^iV) and (^iV) reflected impulses. Accordingly, the transmission line can be represented by an impedance Z in series with a voltage source of magnitude 2^iV.

We are now in a position to monitor the progress of a single pulse as it scatters around the mesh. Since we are dealing with a Cartesian mesh, we can describe the directions of scatter by the compass directions N, S, E, W. Let us consider ^iV_W, which is incident from the west onto a two-dimensional TLM node (x, y) as shown on the left in Fig. 4.4.

As the pulse arrives at the end of its line what it sees in front of it and how it appears are shown on the right. Thus the pulse will undergo scattering according to eqn (4.6). In this case Z_T is equal to $Z/3$, the three impedances seen in parallel by the incoming pulse, and thus $\rho = -1/2$. This means that a pulse of magnitude $-^iV_W/2$ is returned down the incoming transmission line.

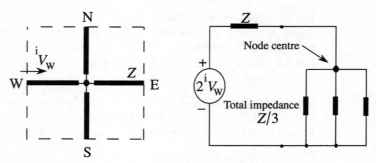

Fig. 4.4 Lossless TLM node and its Thévenin equivalent.

The remainder of the signal is transmitted into the other arms. Pulses incident from arms N, S, and E are simultaneously incident and undergo scattering. They also contribute to the voltage at the node centre which can be represented by the superposition of all voltages, giving the overall voltage at the node at the kth time interval:

$$_k\phi(x, y) = \frac{1}{2}(^iV_N + {}^iV_S + {}^iV_E + {}^iV_W).\tag{4.7}$$

The pulse which is scattered back in any direction is the sum of what is reflected and what is transmitted from all other arms (Fig. 4.5). Thus

$$^sV_W = \rho^iV_W + \tau(^iV_N + {}^iV_S + {}^iV_E).\tag{4.8}$$

There are similar equations for ^sV_N, ^sV_S, and ^sV_E and the entire scattering process can be expressed in matrix form:

$$_k\begin{bmatrix} ^sV_N \\ ^sV_S \\ ^sV_E \\ ^sV_W \end{bmatrix} = \frac{1}{2}\begin{bmatrix} -1 & 1 & 1 & 1 \\ 1 & -1 & 1 & 1 \\ 1 & 1 & -1 & 1 \\ 1 & 1 & 1 & -1 \end{bmatrix}_k\begin{bmatrix} ^iV_N \\ ^iV_S \\ ^iV_E \\ ^iV_W \end{bmatrix}.\tag{4.9}$$

Each pulse travels the discretization distance Δx during the discretisation time Δt after which it becomes an incident pulse at an adjacent node. The connections to other nodes as seen at node (x, y) can be expressed in terms of space and time step, $k + 1$ as

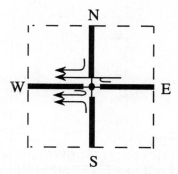

Fig. 4.5 The impulse scattered to the west is the sum of four components.

$$_{k+1}{}^{i}V_N\,(x, y) = {}_k{}^{S}V_S\,(x, y + 1)$$
$$_{k+1}{}^{i}V_S\,(x, y) = {}_k{}^{S}V_N\,(x, y - 1)$$
$$_{k+1}{}^{i}V_E\,(x, y) = {}_k{}^{S}V_W(x + 1, y)$$
$$_{k+1}{}^{i}V_W\,(x, y) = {}_k{}^{S}V_E\,(x - 1, y). \tag{4.10}$$

The repeated application of eqns (4.7), (4.9), and (4.10) for every time step allows us to describe the propagation of the single initial impulse, $^{i}V_W$, throughout the mesh as a function of time.

A simple algorithm

The concept of repeated scatter, connect, and sum can be extended to more general cases by substituting any initial input as required. We start with a simple two-dimensional array which can be displayed on a computer screen. Initially, all values except one are set to zero. The exception is $^{i}V_N(i, j) = 1024$, where (i, j) is approximately at the mid-point of the screen. Equations (4.9) and (4.108) are used to scatter and connect over 8 repeats while the space begins to fill with pulses. At each time step eqn (4.7) is used to calculate the nodal potential. Specific values have been chosen so that a simple 'C' program, without system-dependent graphics, can be written. The reader could add initial excitations of 1024 for $^{i}V_S(i,j)$, $^{i}V_E(i,j)$ and $^{i}V_W(i,j)$.

```c
#include  "stdio.h"
#include  "conio.h"                           /* Borland Turbo C */

#define SIZE 21                               /* elements 0 to SIZE-1 */

void  main(void)
{
int  vtotal[SIZE][SIZE]={0},
   vin[SIZE][SIZE]={0},vis[SIZE][SIZE]={0},vie[SIZE][SIZE]={0},
   viw[SIZE][SIZE]={0},vsn[SIZE][SIZE]={0},vss[SIZE][SIZE]={0},
   vse[SIZE][SIZE]={0},vsw[SIZE][SIZE]={0};

int counter, count1, count2;

vin[SIZE/2][SIZE/2]=1024;

for(counter=1;  counter  <=8;counter++)        /* eight iterations */
   {

   for (count1=0;count1 < SIZE; count1++)       /* scatter */
      {
      for (count2=0;count2 < SIZE; count2++)
         {
         vsn[count1][count2]=(-vin[count1][count2]  +vis[count1][count2]
                        +vie[count1][count2]  +viw[count1][count2])/2;
         vss[count1][count2]=(+vin[count1][count2]  -vis[count1][count2]
                        +vie[count1][count2]  +viw[count1][count2])/2;
         vse[count1][count2]=(+vin[count1][count2]  +vis[count1][count2]
                        -vie[count1][count2]  +viw[count1][count2])/2;
         vsw[count1][count2]=(+vin[count1][count2]  +vis[count1][count2]
                        +vie[count1][count2]  -viw[count1][count2])/2;
         }
      }

   for (count1=1;count1 < SIZE-1; count1++)       /* connect */
      {
      for (count2=1;count2 < SIZE-1;  count2++)
         {
         vin[count1][count2]=+vss[count1][count2+1];
         vis[count1][count2]=+vsn[count1][count2-1];
```

```
    vie[count1][count2]=+vsw[count1+1][count2];
    viw[count1][count2]=+vse[count1-1][count2];
    }
  }

for (count1=0;count1 < SIZE; count1++)        /* total */
  {
  for (count2=0;count2 < SIZE; count2++)
    {
    vtotal[count1][count2]=(+vin[count1][count2]  +vis[count1][count2]
                      +vie[count1][count2]  +viw[count1][count2])/2;
    }
  }

/* print the centre of the array only */
   printf("\n\n");
   for (count2=SIZE/2+8;count2  >=  SIZE/2-8;count2--)
     {
     printf("\n");
     for (count1=SIZE/2-8;count1  <=  SIZE/2+8;  count1++)
       {
       printf("%4d",vtotal[count1][count2]);
       }
     /* pause after each line printed (Turbo C) */
     getch();
     }
   }
}
```

An alternative view of the output for a slightly larger problem with four initial inputs to a point is shown in Fig. 4.6. This looks quite impressive, but careful examination reveals some curious behaviour. This is due to some important aspects of TLM which we will now consider.

Dispersion characteristics

The first point to note is that in order for a pulse to travel diagonally (a distance of $\sqrt{2}\Delta x$) it must move there during two iterations. Thus the diagonal velocity on a Cartesian mesh is $\sqrt{2}\Delta x/2\Delta t$ or (free space velocity)$\sqrt{2}$. This is solely a feature of the fact that we have discretized movement through space.

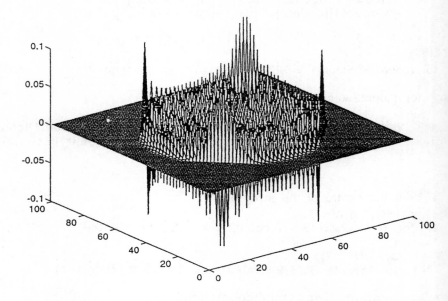

Fig. 4.6 Propagation due to a single excitation at the centre of the two-dimensional mesh. Note that the results should be a uniform circle.

Next, it should be remembered that, according to Fourier principles, any pulse can be constructed out of a spectrum of waves of different wavelength. There is a process termed dispersion which accounts for the fact that signals of different wavelength travel through the mesh at different velocities. The relationship between frequency and the velocity of propagation of a wave on a discretized transmission line network has been deduced and the results for a diagonal plane wave are summarized in Fig. 4.7. This shows that the dispersion curve intersects the vertical axis at the point $1/\sqrt{2}$ and is the maximum velocity which the wave can have on a cartesian mesh. The horizontal axis is given in terms of $\Delta x/\lambda$, the mesh discretization and the signal wavelength (you can use the relationships, $f = v/\lambda$ and $v = \Delta x/\Delta t$ to show that $\Delta x/\lambda = f\Delta t$). The propagation curve intersects the horizontal axis at $\Delta x/\lambda = 0.25$ which defines a *cut-off* condition.

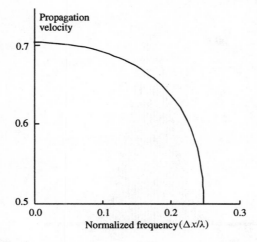

Fig. 4.7 Dispersion curve; the variation of wave velocity as a function of wavelength and discretization (expressed as $\Delta x/\lambda$). Note that the wave velocity is zero at $\Delta x/\lambda = 0.25$.

Now, an impulse signal contains all frequencies. The initial (impulsive) excitation of 1024 in our example therefore contains frequencies which are at or beyond cut-off. On this basis the results in Fig. 4.6 are not unexpected.

For our results to be physically meaningful we need to ensure that all the frequencies which arise in the simulaton are less than cut-off and that (if possible) they move at about the same velocity. Experience shows that this requires $\Delta x/\lambda$ to be less than or equal to 0.1, and it is known that this condition can be approached using an initial excitation with a Gaussian profile. This is shown in Fig. 4.8 where a point at the centre of a mesh was excited at successive time steps with a series of pulses. The propagating waveform which is observed is much closer to what might have been expected and tends to confirm the hypothesis that waves beyond cut-off can lead to erroneous behaviour.

Boundaries

Many physical problems have natural boundaries which conveniently define a limit for the modelling space. Other problems may be partially bounded or unbounded, e.g. sound propagation from a trumpet bell, radio propagation from a monopole on a ground-plane, radio propagation from a dipole in open space. Memory constraints on computers may limit our ability to model these open-bounded problems, but even before we start to ponder such questions we should ask how TLM treats any boundary. In fact it does this by using the same definitions as would be used in the mathematical analysis of electromagnetic waves.

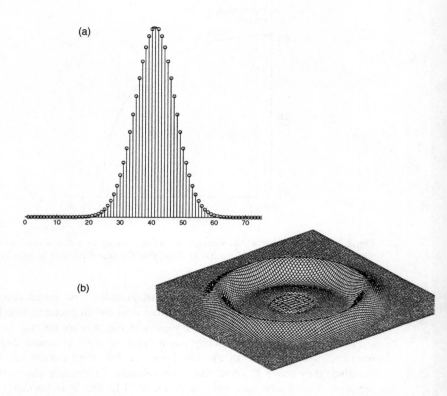

Fig. 4.8 (a) Approximation to a Gaussian excitation which was used to obtain the Huygens propagation shown in (b). Note the small perturbations at the centre of the propagating wavefront. These are due to the fact that the components in our excitation are not all above cut-off.

1. Eqn (4.6) shows that if there is an open-circuit termination, then $Z_T = \infty$. This means that pulses which arrive at a boundary are reflected in phase, because $\rho = 1$.

2. If on the other hand there is a short circuit then $Z_T = 0$. The reflection coefficient, $\rho = -1$ and any pulse incident on a boundary will be reflected in anti-phase.

3. In the situation where $Z_T = Z$ then $\rho = 0$. This is called a *matched load* boundary condition and will feature in further discussion

The Huygens propagation shown in Fig. 4.8 had not yet reached the edges of the mesh and therefore was unaware of the presence of any boundaries. In fact two different classes of boundaries were defined for illustration purposes

Fig. 4.9 Huygens propagation following reflection from one short-circuit (lower right) and one open-circuit (lower left) boundary.

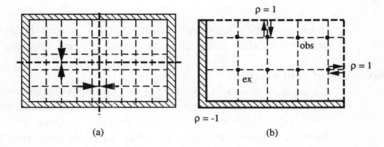

(a)   (b)

Fig. 4.10 (a) A section of discretized electromagnetic waveguide showing the passage of impulses across the two orthogonal symmetry axes. (b) Reduction of this model by means of symmetry. Pulses arriving at the open-circuit boundaries will be reflected exactly as in (a). Pulses approaching the conductor walls will experience a short-circuit boundary.

and when the programme was allowed to run on we begin to see the effects of reflection at the open-circuit (left side) and at the short-circuit (right side) boundaries and this is shown in Fig. 4.9.

The concept of open-circuit boundaries can also be used to reduce the size of the problem which needs to be computed. Setting $\rho = 1$ does not distinguish between a pulse which is reflected in phase at a boundary and one which passes through a symmetry axis.

The structure which is shown in Fig. 4.10 represents the first problem to be analysed using TLM (by Johns and Beurle [4.2] in 1971). The node marked 'ex' is the point at which an input excitation was applied. This was allowed to propagate through the mesh using eqns (4.9), (4.10). The values of impulses which arrived at the observation point ('obs' in the figure) were collected during 100 or more iterations and were then subjected to a discrete Fourier

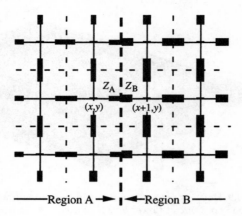

Fig. 4.11 A material discontinuity half-way between two sets of nodes.

transform (DFT) (see Chapter 8 for frequency domain details). The frequency components at the observation point were thereby estimated.

Maxwell field components

On the basis that the value at the point (x, y) is a voltage it is possible to derive some of the field components for Maxwell's equations. For example $[^{i}V_{N}(x,y) - {}^{i}V_{S}(x,y)]/Z\Delta y$ represents $H_{y}(x,y)$. Once we have determined the resonances in the structure (Fig. 4.10) we can drive the input at the excitation point at a fixed frequency. It is therefore possible to obtain the field components equivalent to any of the structure's natural resonances. The convenience that V and I are isomorphic with E and H is one of the reasons why the TLM technique has found particular favour with high-frequency electromagnetics specialists.

Changes in material properties: the use of internodal reflection

In many cases we may be required to treat situations where there is a change in material properties. The passage of light from air into glass (refraction) is a typical example. Such changes can be incorporated into a TLM model:

• by using changes in capacitance to represent changes in dielectric permittivity

• by using changes in inductance to represent changes in permeability.

As a consequence of these changes, additional process steps will be needed in the TLM algorithm. At the start of an iteration pulses will leave nodes (x, y) and $(x + 1, y)$ as shown in Fig. 4.11, their magnitude being determined by eqn (4.9). They will each travel a distance $\Delta x/2$ after which they will encounter

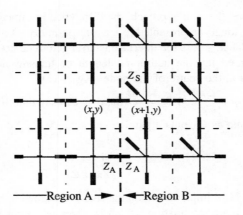

Fig. 4.12 Differences in material properties loaded into stubs.

a change in impedance which will cause additional scattering. In general terms a pulse moving from its source in B to an adjacent node in A will experience a reflection if Z_A and Z_B are different. The reflection coefficient is given by:

$$\rho_{B \to A} = \frac{Z_A - Z_B}{Z_A + Z_B}. \quad (4.11)$$

The pulse arriving at B from A will experience a complementary reflection coefficient

$$\rho_{A \to B} = \frac{Z_B - Z_A}{Z_B + Z_A}.$$

The transmission coefficients are $\tau_{B \to A} = 1 - \rho_{B \to A}$ and $\tau_{A \to B} = 1 - \rho_{A \to B}$.
The connection process across the boundary now becomes:

$$\begin{aligned}
_{k+1}{}^{i}V_E(x, y) &= \rho_{A \to B} \, _{k}^{S}V_E(x, y) + \tau_{B \to A} \, _{k}^{S}V_W(x + 1, y) \\
_{k+1}{}^{i}V_W(x + 1, y) &= \rho_{B \to A} \, _{k}^{S}V_W(x + 1, y) + \tau_{A \to B} \, _{k}^{S}V_E(x, y)
\end{aligned} \quad (4.12)$$

These alterations are only relevant to pulses crossing the boundary between the two materials. Otherwise, since the impedances across nodes are identical, eqn (4.12) can be expressed similarly to eqn (4.10).

Changes in material properties: the use of stubs

There is a totally different way of approaching the problem of material discontinuities, which requires a slightly more complicated scattering matrix (eqn (4.9)) but which has no requirement for an intermediate scattering of the form shown in eqn (4.12). This uses the concept of stubs, borrowed from microwave engineering.

Using the stub approach, the difference in impedance between two regions is no longer represented by a discontinuity at the interface. Instead, the impedance across the interface is arranged to be constant and the difference is subsumed into an additional length of transmission line (a stub) which is placed at the centre of the node (Fig. 4.12).

If $Z_A = \dfrac{\Delta t}{C_A}$ and $Z_B = \dfrac{\Delta t}{C_B}$ then we can say that the difference between C_B is the same as C_A plus a parallel capacitance C_S. So, if $C_B = C_A + C_S$, we can write

$$\frac{1}{Z_B} = \frac{1}{Z_A} + \frac{1}{Z_{St}}.$$

In the TLM implementation the stub is designed to have a length $\Delta x/2$ with an open-circuit termination, which is effectively a capacitance. A pulse launched into a stub will travel for time $\Delta t/2$, then it will be reflected back in phase and will arrive into the original node as other pulses are coming in from other surrounding nodes.

The connect part of the TLM routine is exactly as per eqn (4.10) with the additions $_{k+1}{}^iV_{St} = {}_k{}^iV_{st}$. It is the scattering process that is different. The impedance of the stub is given by:

$$Z_{St} = \frac{\Delta t/2}{C_S} = \frac{\Delta t}{2\,C_S}. \tag{4.13}$$

The derivation of a scattering matrix appears complicated because the circuit in Fig. 4.4 now has an extra component Z_{St} in the load. Consequently, the scattering between connecting lines must also include scattering into and from the stub. This yields a matrix equation, which is in fact computationally efficient and avoids some of the complexities of the discontinuity approach:

$$\begin{bmatrix} {}^sV_N \\ {}^sV_S \\ {}^sV_E \\ {}^sV_W \\ {}^sV_{St} \end{bmatrix} = \frac{1}{4Z_{st} + Z} S \begin{bmatrix} {}^iV_N \\ {}^iV_S \\ {}^iV_E \\ {}^iV_W \\ {}^iV_{St} \end{bmatrix} \tag{4.14}$$

where

$$S = \begin{pmatrix} -(Z+2Z_{st}) & 2Z_{st} & 2Z_{st} & 2Z_{st} & 2Z \\ 2Z_{st} & -(Z+2Z_{st}) & 2Z_{st} & 2Z_{st} & 2Z \\ 2Z_{st} & 2Z_{st} & -(Z+2Z_{st}) & 2Z_{st} & 2Z \\ 2Z_{st} & 2Z_{st} & 2Z_{st} & -(Z+2Z_{st}) & 2Z \\ 2Z_{st} & 2Z_{st} & 2Z_{st} & 2Z_{st} & (Z - 4Z_{st}) \end{pmatrix}.$$

The stub effectively acts as a temporary storage element which always retains some residue of any pulse that enters it. We can imagine a unit pulse which enters a stub ($Z_{St} = 1$) from a line whose charactersitic impedance is 1. When it arrives back from the open circuit after Δt it sees the transmission line in front of it as two impedances in parallel which is equivalent to a total load of 1/2. A portion of the pulse is therefore scattered back into the stub with a reflection coefficient $(0.5 - 1)/(0.5 + 1) = -1/3$. The remainder enters the line and for the purpose of this analysis disappears. After another full time step the pulse in the stub is again reflected so that the history of the initial pulse that entered the stub is spread over subsequent iterations. Note that 2/3 of any pulse, transmitted back into the line goes left and 2/3 goes right. This means that the total is $-1/3$(pulse magnitude reflected into stub) $+2/3$(magnitude transmitted into line right) $+2/3$(magnitude transmitted into line left) which of course adds to the magnitude of the pulse at the start of the time step. The realization of this in code does not necessarily require the use of high-level computer languages. The algorithm is as follows and can be implemented in most standard spreadsheet packages.

$k = 0$ *refl*ected into stub $= 1$ *trans*mitted into a line $= 0$

$k > 0$ $\text{refl}_k = (-1/3)*\text{refl}_{k-1}$ $\text{trans}_k = (1/2)*(\text{refl}_{k-1} - \text{refl}_k)$

Thus:

A1 $= 1$, A2 $= (-1/3)*$A1   and is used as the fill-down formula

B1 $= 0$, B2 $= (1/2)*($A1 $-$ A2$)$ and is used as the fill-down formula.

4.2.2 Series mesh representation

The series network is an alternative TLM representation which is extensively used for two-dimensional models of electromagnetic problems. It is slightly more difficult to visualize because it cannot be easily represented by the coaxial cables as in Figs 4.2 and 4.3. However if we were to assemble a line of four metal boxes on a surface with each box separated from the next by a small gap then it would be easy to see this as a collection of capacitors in series. We could fill an area of surface with equi-spaced metal boxes. If we were then to connect opposite diagonals to a battery then this two dimensional array of series capacitors would charge up and the rate at which charge built up (dQ/dt otherwise known as current) would depends on the inductance within the conductors. Figure 4.13(a) shows the region where the diagonals of four metal boxes meet. Alternatively this can be viewed as a node point where four series connected transmission lines coincide. In a formal sense this provides the Maxwell components E_x, E_y, and H_z if we assume that the $x - y$ axis is in the plane of the paper. The lumped equivalent circuit is shown in Fig. 4.13(b).

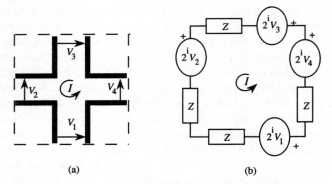

(a)　　　　　　　　　(b)

Fig. 4.13 A TLM series mode and its lumped equivalent circuit.

The current circulating in the node is given by

$$I = \frac{2^i V_1 - 2^i V_2 - 2^i V_3 + 2^i V_4}{4Z} \qquad (4.15)$$

and from this it can be shown that the matrix scattering equation is

$$\begin{bmatrix} {}^sV_1 \\ {}^sV_2 \\ {}^sV_3 \\ {}^sV_4 \end{bmatrix}_k = \frac{1}{2} \begin{bmatrix} 1 & 1 & 1 & -1 \\ 1 & 1 & -1 & 1 \\ 1 & -1 & 1 & 1 \\ -1 & 1 & 1 & 1 \end{bmatrix} \begin{bmatrix} {}^iV_1 \\ {}^iV_2 \\ {}^iV_3 \\ {}^iV_4 \end{bmatrix}_k. \qquad (4.16)$$

4.2.3 Extensions to three dimensions

Three-dimensional electromagnetic models require all six Maxwell components H_x, H_y, H_z, E_x, E_y, and E_z to be calculated. The series and shunt representations presented above each yield three components, and are therefore applicable to one- and two-dimensional problems only. The extension of lossless TLM to three dimensions is almost beyond the scope of this book and is introduced here only for completeness, along with references to the original work.

The expanded node network

The expanded node network [4.3] is a joining together of series and shunt meshes in three-dimensional space as shown in Fig. 4.14.

The series and shunt nodes of the expanded network can be scattered independently of each other, but the pulses scattered by a series node will be incident on a shunt node and vice versa. The primary derived variable at different positions is shown in the figure and for each mesh point two other

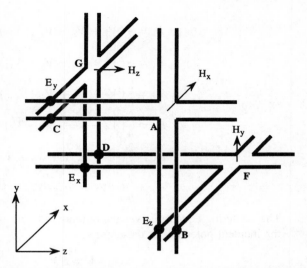

Fig. 4.14 Part of an expanded node network showing the field components as the primary derived variable at different positions. Note that positions A, B, and C are in the same plane and are closest to the viewer. Positions D, F, and G are in the same plane with respect to each other, but are distance Δx beyond A, B, and C.

components can be derived. Thus point A yields H_x, E_y, and E_z. Point B similarly yields H_x, H_y, and E_z. So for any point in three-dimensional space we can derive half of the required field components. So long as the gradients are not too large it is possible to determine the missing values by interpolation. This technique has several disadvantages, the most significant being the difficulty of effectively defining boundary conditions which apply to both meshes.

Condensed TLM nodes

An improvement on the expanded-node method is the condensed-node method, which attempts to define the field quantities within a single cell. Initially development was concerned with obtaining the field quantities at different locations within a cell. This asymetrical approach [4.4] was quickly superceded by the symmetrical condensed node (SCN) approach [4.5]. The interpretation in terms of series and shunt meshes is more difficult in the latter case, but the diagrammatic representation (Fig. 4.15) is beautiful and the scattering matrix is elegant.

The incident and reflected pulses in an SCN appear on the terminals of the transmission lines at ports which are numbered and directed as shown in the figure. If the discretizations in x, y, and z are equal, designated as h, and if the impedances in the three directions are also equal, then the field quantities can be determined as:

$$E_x = \frac{2}{h}({}^iV_1 + {}^iV_2 + {}^iV_9 + {}^iV_{12})$$

$$E_y = \frac{2}{h}({}^iV_3 + {}^iV_4 + {}^iV_8 + {}^iV_{11})$$

$$E_z = \frac{2}{h}({}^iV_5 + {}^iV_6 + {}^iV_7 + {}^iV_{10})$$

$$H_x = \frac{2}{h}({}^iV_4 - {}^iV_5 + {}^iV_7 - {}^iV_8)$$ \hfill (4.17)

$$H_y = \frac{2}{h}(-{}^iV_2 + {}^iV_6 + {}^iV_9 - {}^iV_{10})$$

$$H_z = \frac{2}{h}(-{}^iV_3 + {}^iV_1 + {}^iV_{11} - {}^iV_{12}).$$

The scattering equation gives the outputs from all of the 12 ports in terms of the incident potentials at these ports:

$$^sV = S^iV \hspace{2cm} (4.18)$$

where the scattering matrix is given by

$$S = \frac{1}{2}\begin{bmatrix} 0 & 1 & 1 & 0 & 0 & 0 & 0 & 0 & 1 & 0 & -1 & 0 \\ 1 & 0 & 0 & 0 & 0 & 1 & 0 & 0 & 0 & -1 & 0 & 1 \\ 1 & 0 & 0 & 1 & 0 & 0 & 0 & 1 & 0 & 0 & 0 & -1 \\ 0 & 0 & 1 & 0 & 1 & 0 & -1 & 0 & 0 & 0 & 1 & 0 \\ 0 & 0 & 0 & 1 & 0 & 1 & 0 & -1 & 0 & 1 & 0 & 0 \\ 0 & 1 & 0 & 0 & 1 & 0 & 1 & 0 & -1 & 0 & 0 & 0 \\ 0 & 0 & 0 & -1 & 0 & 1 & 0 & 1 & 0 & 1 & 0 & 0 \\ 0 & 0 & 1 & 0 & -1 & 0 & 1 & 0 & 0 & 0 & 1 & 0 \\ 1 & 0 & 0 & 0 & 0 & -1 & 0 & 0 & 0 & 1 & 0 & 1 \\ 0 & -1 & 0 & 0 & 1 & 0 & 1 & 0 & 1 & 0 & 0 & 0 \\ -1 & 0 & 0 & 1 & 0 & 0 & 0 & 1 & 0 & 0 & 0 & 1 \\ 0 & 1 & -1 & 0 & 0 & 0 & 0 & 0 & 1 & 0 & 1 & 0 \end{bmatrix}$$

The SCN is already a very popular component in TLM modelling and is a key feature in commercial packages such as STRIPES.

4.2.4 Examples of lossless propagation using shunt meshes

(a) Wave propagation through a lens

This example demonstrates an application of lossless TLM. A plane wave is driven at a fixed frequency and is allowed to propagate towards a region of different permittivity (modelled using stubs) which has the geometry of a convex lens. The effects of standing waves inside the lens as well as focusing beyond the lens are clearly visible (see Fig. 4.16).

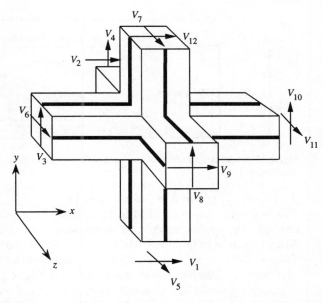

Fig. 4.15 A single symmetric condensed node showing the twelve potentials which occur in the scattering matrix.

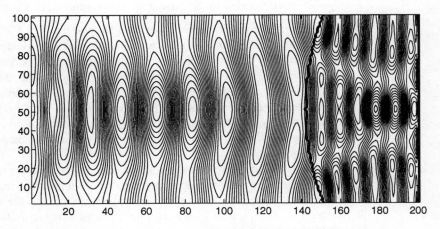

Fig. 4.16 Focusing effect of a curved boundary between two media. The source is driven at a constant frequency.

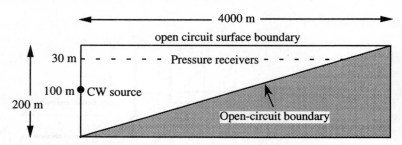

Fig. 4.17 Buckingham and Tolstoy wedge for acoustic propagation in a littoral environment.

(b) The Buckingham and Tolstoy wedge

This is an application of TLM to an artificial benchmark to test underwater acoustic propagation models. It demonstrates some of the computational difficulties of large problems. The test piece, which is due to Buckingham and Tolstoy [4.6] is a section of the sea adjacent to the shore (a littoral environment). The dimensions are as shown in Fig. 4.17.

Ideally we would like to model the propagation of sound of any frequency but inspection of the dispersion curve (Fig. 4.7) indicates that computational problems may arise. For uniform propagation we need a ratio of discretization to wavelength $(\Delta x/\lambda)$ less than 0.1. The velocity of sound in water is approximately $1500~\text{ms}^{-1}$. Therefore, at a frequency of 1.5 kHz there is a wavelength of 1 m. This means that the discretization must be not more than 10 cm. For the problem cited above this would require a total of $\dfrac{4000 \times 100}{2} \times 100$ nodes to be solved at every time step. Clearly this is beyond the scope of much current computational power.

An excitation of 25 Hz was used in this case, and this yielded a more tractable problem. The triangle of water was divided into 1000 nodes along the horizontal and 50 at the deepest point on the vertical. The wedge was simulated by increasing the height of the sea-bed $(\rho = 1)$ boundary by one node for every 20 horizontal nodes.

The results are plotted as pressure intensity as a function of range and are shown in Fig. 4.18. They are in very close agreement with predictions using other techniques.

A particularly obvious feature in the figure is the process of 'mode stripping'. Where the water is deepest it is capable of supporting 7 maxima. As we move towards the shore the number of supportable maxima (or antinodes) is reduced to 5, then 3, then 1. A hydrophone placed in water shallower than this would be unlikely to receive any substantial signal from the sound source.

As this problem was presented here, only signals which travelled from the source to the shore were considered. We have effectively ignored any effect

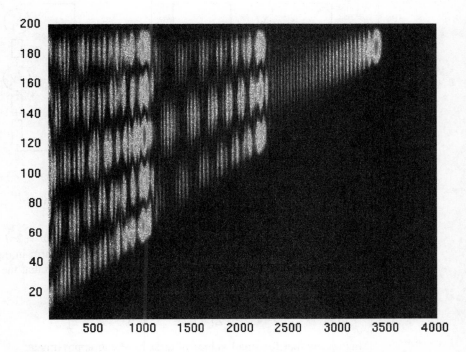

Fig. 4.18 Regions of constant pressure between the two boundaries in a Buckingham and Tolstoy wedge simulation.

due to signals which propagated in the opposite direction. In reality, there would be back scatter from such signals, and in an accurate model these would have to be taken into account.

(c) Modelling electrical filters and circuits

In conventional electrical theory the impedance of a capacitor is given by $Z_C = 1/j\omega C$. There is a slightly different situation in microwave engineering where we are frequently interested in the apparent impedance, Z_{obs}, as observed from one point on a transmission line due to a termination at some other point. The impedance sometimes appears to represent a capacitance and at other times an inductance, this depends on the signal frequency and on the observation point as explained below.

If a transmission line of length $\Delta x/2$ is terminated by an open circuit the apparent impedance is given by [4.7]:

$$Z_{obs} = \frac{Z}{j \tan(\omega \Delta t/2)}. \tag{4.19}$$

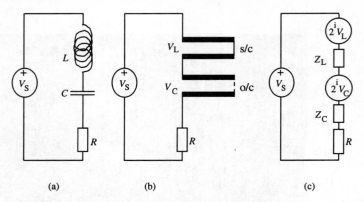

(a) (b) (c)

Fig. 4.19 An RLC circuit with its TLM and lumped circuit equivalents.

This represents a capacitance for $0 < \omega\Delta t/2 < \pi/2$ where $\omega = 2\pi f$ (f is the frequency). Within this range $Z_C = Z_{obs}$. The $\tan(\omega\Delta t/2)$ term in eqn (4.19) can be expanded using a Maclaurin series (see Appendix) so that the equality can be rewritten as:

$$j\omega C = \frac{j}{Z}\left[\omega\Delta t/2 + \frac{(\omega\Delta t/2)^3}{3} + \dots\right]. \tag{4.20}$$

This means that for small values of $\omega\Delta t/2$ we can approximate:

$$C = \frac{\Delta t}{2Z} \tag{4.21}$$

which is consistent with our stub models developed earlier.

Subject to the condition of small values of $\omega\Delta t/2$, an open-circuit terminated transmission line can be used in place of a capacitor in an electronic circuit. Similarly, we can replace inductors by short-circuit terminated transmission lines ($L = Z\,\Delta t$).

Thus we can model a simple RLC circuit, like that shown in Fig. 4.19. The circuit in Fig. 4.19(c) can be used to predict the current, $_kI$:

$$_kI = \frac{V_S - 2_k{}^iV_C - 2_k{}^iV_L}{R + Z_C + Z_L}. \tag{4.22}$$

This is now used to calculate the voltages

$$_kV_C = 2_k{}^iV_C + {}_kIZ_C$$
$$_kV_L = 2_k{}^iV_L + {}_kIZ_L \tag{4.23}$$
$$_kV_R = {}_kIR.$$

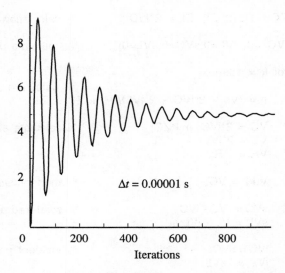

Fig. 4.20 Unscaled capacitor voltage as a function of iteration derived from the Matlab implementation of LRC circuit.

The voltages which are scattered into the respective transmission lines are

$$_k^s V_C = {}_k V_C - {}_k^i V_C$$
$$_k^s V_L = {}_k V_L - {}_k^i V_L. \tag{4.24}$$

These are reflected at the terminations and arrive back at the next iteration as:

$$_{k+1}^i V_C = {}_k^s V_C$$
$$_{k+1}^i V_L = -{}_k^s V_L. \tag{4.25}$$

These values are then fed back into eqn (4.22) for the next time step.

Equations (4.22)–(4.25) are all that is required to predict the circuit current, $_k I$ or any of the measurable voltages $_k V_C$, $_k V_L$, $_k V_R$ at each discrete instant in time $k\Delta t$. An example of the application of this circuit for $L = 10$ mH, $C = 1$ μF and $R = 10\Omega$U with $\Delta t = 10$ μS is given in Matlab code below. The predicted frequency (from $1/(2\pi\sqrt{LC})$ is 1592 Hz which compares very favourably with 1587 Hz derived from the graphical ouput of the program shown in Fig. 4.20.

- -

```
%  TLM simulation of LRC circuit

L=1e-2;C=1e-6;R=10;Vs=5;
Dt=1e-5;kmax=1000;                   % data input
```

```
ZC= Dt/(2*C);  ZL= 2*L/Dt;          % calculation of impedances

iVC =0; iVL=0;sVC=0;sVL=0;          % initialize pulses

for k = 1:kmax
                                    % calculate circuit current
    I = (Vs - 2*iVC - 2*iVL)/(R + ZC + ZL);

    VC = 2*iVC  + I*ZC;             % calculate element voltages
    VL = 2*iVL  + I*ZL;
    VR = I*R;

    V(k) = VC;                      % monitor capacitor voltage

    sVC = VC - iVC;                 % scattered pulses
    sVL = VL - iVL;

    iVC = sVC;                      % incident pulses for next iteration
    iVL = -sVL;
end

k=1:kmax;  plot(k,V(k))
```

- -

Applications of these basic ideas to modelling of more complex electric circuits can be found in references such as [4.8].

(d) TLM networks with lossy components

The ideas in the last example can now be fed back into the general theory so that the scope of application can be broadened. Let us imagine two lossless TLM networks which are linked by resistors as shown in Fig. 4.21. In real life this might be a cheap piece of coaxial cable whose conductor has a resistance R and whose insulator has a leakage which is represented by the resistance r.

In order to treat this effectively, we need only consider the section between the two networks and the Thévenin equivalent is given in Fig. 4.22.

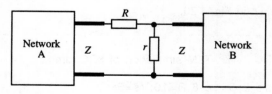

Fig. 4.21 A lossy linkage between two transmission line networks.

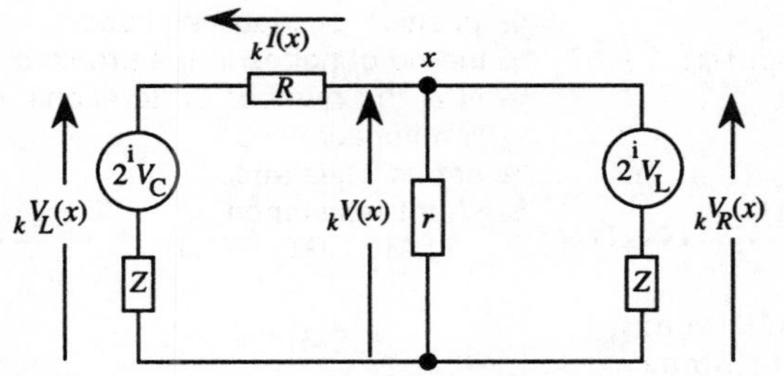

Fig. 4.22 Thévenin equivalent of the lossy linkage between two lossless networks.

The point '*x*' is merely marked for convenience. The potential at *x* at time *k* is given by:

$$_kV(x) = \frac{\frac{2^iV_L(x)}{R+Z} + \frac{2^iV_R(x)}{Z}}{\frac{1}{Z} + \frac{1}{R+Z} + \frac{1}{r}}.$$

The current is given by:

$$_kI(x) = \frac{_kV(x) - 2^iV_L(x)}{R+Z}.$$

The voltage on the right side is

$$_kV_R(x) = {_kV(x)}$$

and on the left side is

$$_kV_L(x) = 2_k{^iV_L(x)} + {_kI(x)Z}.$$

In each case $kV_R(x)$ and $_kV_L(x)$ are the summation of incident and scattered pulses. The incident pulses are known and thus the pulses scattered back into the networks are:

$$_k{^sV_L(x)} = {_kV_L(x)} - {_k{^iV_L(x)}}$$
$$_k{^sV_R(x)} = {_kV_R(x)} - {_k{^iV_R(x)}}$$

These become incident within the respective networks at the next iterations and thus the process continues. The implementation of these ideas is shown in a piece of Matlab code below:

```
% TLM algorithm for an asymmetrical lossy linkaged between two identical lossless
networks
%    ********************    INPUT    PARAMETERS    ***********************
```

```
input=3;                          % position of input excitation
nmax = 2*input-1;                 % the no of nodes in the problem
N=100;                            % N is the number of iterations
Z=1;                              % line impedances
R=1;                              % series resistance
r=1/R;                            % shunt resistance
%**********************    TLM   ROUTINE    **************************
% initialize
   vir=zeros(1,nmax);
   vil=zeros(1,nmax);
   vsr=zeros(1,nmax);
   vsl=zeros(1,nmax);
   vr=zeros(1,nmax);
   vl=zeros(1,nmax);
   current=zeros(1,nmax);
% input
   vir(input)  =  1000;
   vil(input)  =  1000;
%the iteration loop
for k=1:N
  %summation of incident pulses
   vtotal = vil +vir;
  %corrections for the lossy linkage
   vtotal(input)=
((2*vil(input))/(R+Z)+(2*vir(input))/Z)/((1/Z)+(1/(R+Z))+(1/r));
   current(input)=  (vtotal(input)  -  2*vil(input))/(R+Z);
   vr(input)=  vtotal(input);
   vl(input)=  2*vil(input)  +  current(input)*Z;
  %scatter
   vsl=vir;
   vsr=vil;
  %corrections for the lossy linkage
   vsl(input)  =  vl(input)  -  vil(input);
   vsr(input)  =  vr(input)  -  vir(input);
  % connect
   for  j=2:nmax
       vil(j)  =  vsr(j-1);
   end
   for  j=1:nmax-1
       vir(j)  =  vsl(j+1);
   end
  % apply boundary conditions
   vil(1)  =  vsl(1);
   vir(nmax)=  vsr(nmax);
  % display the nodal voltages after each iteration
   disp(vtotal)
end
```

4.3 Lossy TLM models

4.3.1 Lumped and distributed networks

It is possible to develop a TLM model with lossy components as in the last example. However, while lossless TLM is generally concerned with vector quantities, we now move on to consider an approach which is more usually applied to scalars such as temperature, pressure, or concentration. In this form of modelling there is substantial distributed resistance and the timescales for change are so large that the wave propagation term in eqn (4.3) can be effectively ignored so that the telegraphers' equation reduces to a diffusion equation. This is the basis of lossy TLM modelling, which is in many ways less restrictive than lossless TLM. As most published work up to this time has been concerned with shunt TLM, the treatment which we present here is restricted to that approach.

The circuit in Fig. 4.23 has a mechanical equivalent involving springs and dashpots (shock absorbers). Either can be effectively used to model thermal conduction and mass diffusion. The electrical analysis can be formulated using a finite-difference approach (Chapter 3) but it suffers from one major drawback: ideal resistors and capacitors do not possess the time delays that are inherent in real world diffusion processes.

The same circuits can be replaced by networks of transmission lines and resistors where time is a relevant parameter. Starting from the expressions in eqn (4.4) and using the fact that velocity $v = \Delta x/\Delta t$ we can show that

$$C = \frac{\Delta t}{Z}. \tag{4.26}$$

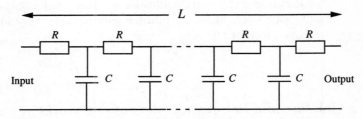

Fig. 4.23 One-dimensional RC ladder network for modelling heat and mass transfer.

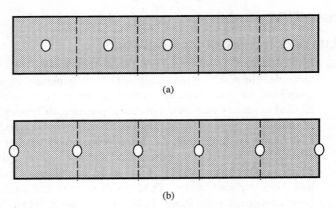

Fig. 4.24 (a) Discretization and positioning of observation points most frequently used in TLM algorithms. (b) Discretization and positioning of observation points most frequently used in finite-difference models.

So, an RC network of length L would be replaced by n lengths of transmission line each separated by a resistor. The transit time for each line section is Δt.

An RC network models heat flow by letting C equal the thermal capacitance of a discrete volume of material. This is of course a function of the material specific heat and density.

$$C = \text{(density) (specific heat) (volume)} = dC_p A \Delta x = C_d \Delta x. \qquad (4.27)$$

There is also a direct relationship between R and the thermal material resistance

$$R = \frac{\text{(contact separation)}}{\text{(contact area) (thermal conductivity)}} = \frac{\Delta x}{A\, k_T} = R_d \Delta x \qquad (4.28)$$

where R_d and C_d are distributed properties.

In the case of matter diffusion, the modeller is not normally interested in the individual values of R and C provided that $(R_d C_d)^{-1}$ is constant. This inverse product is identical with the diffusion constant, and it is this relationship which links Fick's second law of diffusion with our approximation to the telegraph equation (eqn (4.5)):

$$\nabla^2 \phi = R_d C_d \partial\phi/\partial t = \frac{1}{D} \partial\phi/\partial t. \qquad (4.29)$$

4.3.2 Link-line and link-resistor discrete networks

Before developing practical algorithms using lossy TLM we need to consider further factors relating to the spatial discretization. The one-dimensional sample of material in Fig. 4.24 is discretized into equal nodes which are shown

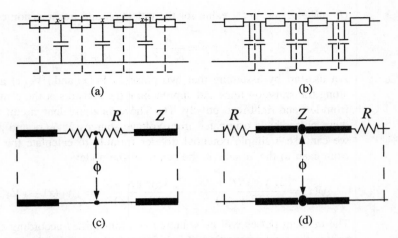

Fig. 4.25 (a) T-network representation of an RC network. (b) Π-network representation of an RC network. (c) Link-line TLM node for diffusion. (d) Link-resistor TLM node for diffusion.

as separated by dashed lines. In case (a) the observation points are placed at the centre of the nodes. In case (b) they are placed at the interfaces. Both are equally valid, but they require slightly different formulations.

It will be noted that the sample in Fig. 4.24 (a) apparently contains one less observation point than the sample in Fig. 4.24 (b). In fact, it contains one more, since presumably the boundaries are already defined. In case (a) the separation between the extreme observation points and the boundaries is $\Delta x/2$, while elsewhere it is Δx. This does not present a problem in TLM but it could increase the complexity in a finite-difference formulation.

The circuit in Fig. 4.23 can be modelled using either a T-network or a Π-network approach. The difference is shown in Fig. 4.25. Exactly the same considerations apply in TLM models for heat flow and diffusion. One node can be separated from the next by means of a transmission line (termed a *link-line representation*) or by means of a resistor (*link-resistor representation*). A single node of each is also shown in Fig. 4.25.

In one-dimension, the link-line and link-resistor treatments are completely equivalent, being simply the translation of the observation point. There are however significant differences in two- and three-dimensional formulations which will be mentioned later.

4.3.3 TLM diffusion algorithms

One-dimensional link-line

A voltage impulse travelling along a transmission line eventually encounters a resistor. Using eqn (4.6) and defining Z_T as the total impedance ahead

$(R + R + Z)$ we get a value for the reflection and transmission coefficients:

$$\rho = \frac{R}{R + Z}; \quad \tau = \frac{Z}{R + Z}. \tag{4.30}$$

Let us start by assuming that two pulses, $_k{}^iV_L(x)$ and $_k{}^iV_R(x)$ are travelling along transmission lines and approaching the resistors at the centre of node x from left and right respectively. The Thévenin equivalent circuit assumes that these pulses have originated from voltage sources $2_k^i V_L(x)$ and $2_k^i V_R(x)$, and we can use a simple potential divider formula to calculate the contribution from each to the voltage at the centre of the node:

$$_k\phi(x) = \frac{2_k{}^iV_L(x)(R + Z)}{(2R + 2Z)} + \frac{2_k{}^iV_R(x)(R + Z)}{(2R + 2Z)} = {}_k{}^iV_L(x) + {}_k{}^iV_R(x). \tag{4.31}$$

The incident pulses will be scattered and transmitted according to eqn (4.30) so that the total summation of pulses moving left and right are:

$$\begin{aligned} _k{}^sV_L(x) &= \rho_k{}^iV_L(x) + \tau_k{}^iV_R(x) \\ _k{}^sV_R(x) &= \tau_k{}^iV_L(x) + \rho_k{}^iV_R(x). \end{aligned} \tag{4.32}$$

Each of the scattered pulses takes time Δt to travel to adjacent nodes, as in eqn (4.10):

$$\begin{aligned} _{k+1}{}^iV_L(x) &= {}_k{}^sV_R(x - 1) \\ _{k+1}{}^iV_R(x) &= {}_k{}^sV_L(x + 1). \end{aligned} \tag{4.33}$$

The repetition of eqns (4.31), (4.32), and (4.33) complete the requirements for a link-line TLM algorithm.

In order to demonstrate some of the more important points up to now we shall take as a running example a 10 cm long bar of copper of 1 cm^2 cross-sectional area. This is to be divided into 10 nodes and we now derive some of the relevant parameters. This will be done using units cm, g, Kelvin (K). The equivalent parameters in units m, kg, Kelvin are shown in brackets — see Table 4.1

The first steps of the link-line algorithm are shown in Table 4.2.

One-dimensional link resistor

The algorithm for a link resistor can also be used to give the potentials at the interface between nodes, since it is simply the summation of left- and right-going pulses. At the start of an iteration six pulses share three positions, $x-1$, x, and $x + 1$ which are situated at the centre of transmission lines as in Fig. 4.25(d) (which is the TLM equivalent of Fig. 4.25(a)). The pulse which is at $x-1$ travelling left is no longer relevant to node x. The same applies to the

Table 4.1 10 cm copper bar (part 1): the TLM parameters

density (d)	$= 8.92$ g/cm^3 (8920 kgm^{-3})
specific heat (C_p)	$= 0.384$ J/g/K (384 Jkg^{-1} K^{-1})
thermal conductivity (k_T)	$= 4.01$ W/cm/K (4010 Wm^{-1}K^{-1})
Diffusivity, $k_T/(d\ C_p)$	$= 1.169$ cm^2/s (1.169 $\times$ 10^{-4} m^{2}s^{-1})

Since $\Delta x = 1$ cm (0.1 m) and $A = 1$ cm^2 (10^{-4} m^2)
Capacitance, $C\quad = d\Delta xA\ C_p = 3.43$
Resistance, $R\qquad = \Delta x/k_T\ A\ = 0.249$
Time constant, $RC = 0.854$ s

Results will only be accurate if Δt is significantly less than RC. We will have $\Delta t = 0.01$ s.

Impedance, $Z\qquad = \Delta t/C\quad = 0.002915$
Reflection, $\rho\qquad = 0.98843$
Transmission, $\tau\quad = 0.01157$

Table 4.2 10 cm copper bar (part 2a): first steps in a link-line algorithm

Let us assume that the initial state in the copper rod is
$$_0{}^iV_L(5) = 100, \quad _0{}^iV_R(5) = 0$$
then the scattered pulses are zero except at $x = 5$:
$$_0{}^sV_L(5) = 0.98843 * 100 + 0.01157 * 0$$
$$_0{}^sV_R(5) = 0.01157 * 100 + 0.98843 * 0.$$
The incident pulses at time 1 are
$$_1{}^iV_L(5) = 0,$$
$$_1{}^iV_R(5) = 0,$$
and elsewhere except
$$_1{}^iV_R(4) = _0{}^sV_L(5) = 98.843$$
$$_1{}^iV_L(6) = _0{}^sV_R(5) = 1.157.$$
Therefore
$$_k\phi(4) = 98.843, \quad _k\phi(5) = 0, \quad _k\phi(6) = 1.157.$$

one which is travelling right from $x+1$. The other four pulses travel for time $\Delta t/2$ before they are scattered at the resistors. They then become incident on x from left and right as

$$_{k+1}{}^iV_L(x) = \rho_kV_L(x) + \tau_kV_R(x-1)$$
$$_{k+1}{}^iV_R(x) = \rho_kV_R(x) + \tau_kV_L(x+1).$$

(4.34)

The pulses arrive simultaneously at x and they sum to give the instantaneous potential

Table 4.3 10 cm copper bar (part 2b): first steps in a link-resistor algorithm

---

Let us assume that the initial state in the copper rod is
$$_0{}^iV_L(5) = 100, \quad _0{}^iV_R(5) = 0.$$
As the pulses pass the node centres
$$_0\phi(4) = 0, \quad _0\phi(5) = 100, \quad _0\phi(6) = 0.$$
After the pulses pass the node centres their identities change
$$_0V_L(5) = 0, \quad _0V_R(5) = 100.$$
These pulses are then scattered at the resistors to become incident
$$\begin{aligned}
_1{}^iV_L(5) &= 0.98843 * 0 + 0.01157 * 0 &&= 0 \\
_1{}^iV_R(5) &= 0.98843 * 100 + 0.01157 * 0 &&= 98.843 \\
_1{}^iV_L(6) &= 0.98843 * 0 + 0.01157 * 100 &&= 1.157 \\
_1{}^iV_R(6) &= 0.98843 * 0 + 0.01157 * 0 &&= 0.
\end{aligned}$$
The summations are then
$$_1\phi(5) = 98.843, \quad _1\phi(6) = 1.157.$$
After the pulses pass the node centres their identities change
$$_1V_L(5) = 98.8430, \quad _1V_R(5) = 0$$
$$_1V_L(6) = 0, \quad _1V_R(6) = 1.157$$
which are subsequently scattered.

$$_k\phi(x) = {}_{k+1}{}^iV_L(x) + {}_{k+1}{}^iV_L(x). \tag{4.35}$$

Once the pulses pass on their way after this incidence it is useful to redesignate them for use at the next iteration

$$\begin{aligned}
_{k+1}V_R(x) &= {}_{k+1}{}^iV_L(x) \\
_{k+1}V_L(x) &= {}_{k+1}{}^iV_R(x).
\end{aligned} \tag{4.36}$$

A complete algorithm consists of the repetition of eqns (4.34)–(4.36) for k iteration steps, where $k\Delta t$ is the total time of the simulation. Table 4.3 gives just the first few steps.

Two- and three-dimensional link line

Although the treatment given here is for a two-dimensional formulation, it can be extended to three dimensions without great difficulty. Figure 4.26 shows a node with a pulse incident from the west at the instant before it is reflected.

The impedance which the pulse sees at the discontinuity is a resistor in series with a parallel arrangement of three impedances $(R + Z)$. The reflection coefficient is therefore

$$\rho = \frac{3R + R + Z - 3Z}{3R + R + Z + 3Z} = \frac{2R - Z}{2R + 2Z}.$$

The transmitted component down any arm is τ, where $3\tau = 1 - \rho$.

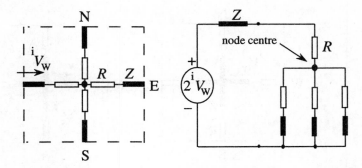

Fig. 4.26 A two-dimensional link-line node and its lumped equivalent circuit with a pulse incident from the west.

The lumped circuit due to a single incident pulse is shown in Fig. 4.26, but we can assume that in a general formulation there are pulses incident from all directions. Therefore the potential at (x, y) is best expressed as the current from each contributor passing through the point divided by the total impedance of the node:

$$
\begin{aligned}
k\phi(x, y) &= \frac{\left[\frac{2_k^i V_N(x, y)}{R+Z} + \frac{2_k^i V_S(x, y)}{R+Z} + \frac{2_k^i V_E(x, y)}{R+Z} + \frac{2_k^i V_W(x, y)}{R+Z}\right]}{\left[\frac{R+Z}{4}\right]} \\
&= \frac{_k^i V_N(x, y) + _k^i V_S(x, y) + _k^i V_E(x, y) + _k^i V_W(x, y)}{2}.
\end{aligned}
\tag{4.37}
$$

The pulse which is scattered from each arm consists of one reflected and three transmitted components as in eqn (4.9):

$$
\begin{pmatrix} {}^s V_N \\ {}^s V_S \\ {}^s V_E \\ {}^s V_W \end{pmatrix}_k = \begin{pmatrix} \rho & \tau & \tau & \tau \\ \tau & \rho & \tau & \tau \\ \tau & \tau & \rho & \tau \\ \tau & \tau & \tau & \rho \end{pmatrix} \begin{pmatrix} {}^i V_N \\ {}^i V_S \\ {}^i V_E \\ {}^i V_W \end{pmatrix}_k.
\tag{4.38}
$$

The connection process is exactly as in eqn (4.10). The repetition of steps (4.36)–(4.39) are all that is required to undertake a two-dimensional diffusion simulation based on the link-line node.

Two-dimensional link-resistor formulation

In a two-dimensional link-resistor node the transmission lines intersect as in Fig. 4.1. This gives rise to scattering at discrete time intervals which are described by eqn (4.7). However the presence of linking resistors gives rise to a second set of four scattering events which occur at the half time intervals. These are shown in Fig. 4.27.

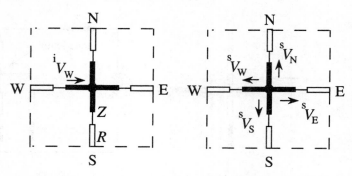

Fig. 4.27 Scattering at $k = 1, 2, 3, \ldots$ and $k = \frac{3}{2}, \frac{5}{2}, \frac{7}{2}, \ldots$ in a link-resistor node.

This additional set of events cause a change in the connection equations:

$$
\begin{aligned}
_{k+1}{}^{i}V_{N}(x, y) &= \rho'^{s}_{k} V_{N}(x, y) + \tau'^{s}_{k} V_{S}(x, y + 1) \\
_{k+1}{}^{i}V_{S}(x, y) &= \rho'^{s}_{k} V_{S}(x, y) + \tau'^{s}_{k} V_{N}(x, y - 1) \\
_{k+1}{}^{i}V_{E}(x, y) &= \rho'^{s}_{k} V_{E}(x, y) + \tau'^{s}_{k} V_{W}(x + 1, y) \\
_{k+1}{}^{i}V_{W}(x, y) &= \rho'^{s}_{k} V_{W}(x, y) + \tau'^{s}_{k} V_{E}(x - 1, y).
\end{aligned}
\tag{4.39}
$$

The nodal potential is as in eqn (4.7).

4.3.4 Inputs

In the previous section we summarized the basis upon which algorithms are built to describe scattering processes in TLM networks, but we did not discuss how pulses might initially be introduced into these networks. In this section we outline various methods of excitation which closely model physical processes. We also cover TLM descriptions of typical boundaries in heat and matter diffusion. The initial treatments are for excitation (pulse injection) into boundary-free (bulk) material. We will consider injection at boundaries after a discussion on boundary definitions.

Single-shot injection into bulk material

This description covers inputs which ultimately lead to Gaussian distributions of matter, heat, or temperature profile. It consists of a voltage source (equivalent to temperature input) or a current source (equivalent to heat input) which is switched across one node point during the first iteration in the simulation. If we were to take the viewpoint of the signal coming from the excitation source then it is clear from Fig. 4.28 that it sees a junction with equal impedance to left and to right.

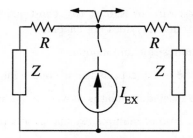

Fig. 4.28 The equal distribution of the output from an excitation source.

Thus the product, $I_{EX}\Delta t$ represents a charge which models the single-shot injection of impurities into a medium. The current divides into equal quantities moving to left and to right. The subsequent diffusion of these current pulses could be monitored using TLM. Now the introductory treatment that has been given in this chapter has been expressed entirely in terms of voltage. This is not a problem, since a current $I_{EX}/2$ passing through $R + Z$ on the left will give rise to a voltage ${}^{i}V_L$. Similarly there will be a voltage ${}^{i}V_R$ moving to the right.

In terms of matter diffusion we could say that if 1000 particles were injected at the source then the initial conditions would be: ${}_{0}^{i}V_L = 1000/2$ and ${}_{0}^{i}V_R = 1000/2$. This would be sufficient to initiate the diffusion algorithm. We can then continue our running example of the copper bar by implementing a 10w continuous heat input and this is demonstrated in Table 4.4.

At this point it is worth considering a peculiarity of the link-line algorithm which the curious reader may have noted when we gave the running example for one-dimensional propagation following a single-shot input. At time $k = 0$ an excitation at x gives rise to pulses travelling left and right. They take time Δt to reach adjacent nodes, so that at $k = 1$ the values of $\phi(x - 1)$ and $\phi(x + 1)$ are defined but $\phi(x)$ is not. Similarly, the pulses scattered at $(x + 1)$ and $(x - 1)$ scatter so that at $k = 2$ the values $\phi(x)$, $\phi(x - 2)$ and $\phi(x + 2)$ are defined while $\phi(x + 1)$ and $\phi(x - 1)$ are not. This will manifest itself as apparent jumps to zero of any node at alternate timesteps, which is obviously unphysical. Such behaviour is observed in other (non-TLM) techniques and in fact represents singularity.

Table 4.4 10 cm copper bar (part 3): heat input as temperature rise

We will assume that the centre of the copper rod which we are using as a running example is connected to a continuous 10 W heat source. Since $\Delta t = 0.01$ s the energy transferred during one time step is 0.1 J. Thus,
${}_{0}^{s}V_L(5) = {}_{0}^{s}V_R(5) = (0.1/3.43)$ °C and the left and right scattered pulses at node 5 are incremented by $(0.1/3.43)$ °C at every subsequent time step.

It can easily be shown (by constructing diagrams similar to those in Fig. 4.29) that this does not happen:

- in link-resistor formulations

- if some of the capacitance is concentrated at each node centre as a stub

- if the excitation point is moved to the boundary between two nodes

- in the case of multiple excitations at a point.

The diagrammatic proof of these assertions is left as an exercise for the reader.

Multiple and constant injection into bulk material

(1) Constant heat source

A constant rate of heating means that an additional input will be required at every time step. It is possible to arrange the input at any point in a TLM formulation, but it is often most convenient to add a pulse (ΔV) immediately

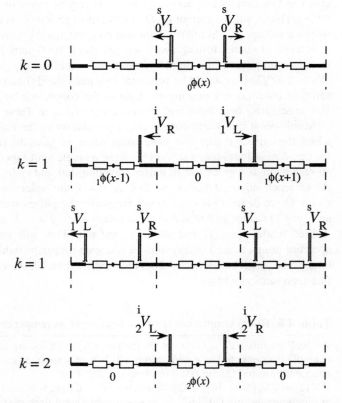

Fig. 4.29 The first three time steps of a single injection showing the anomalous 'jumps to zero' which arise in a link-line representation of diffusion.

after the incident step at each iteration so that the nodal potential is:

$$_{k+1}\phi(x) = _{k+1}{}^{i}V_{L}(x) + _{k+1}{}^{i}V_{R}(x) + \Delta V. \tag{4.40}$$

Half of the heat is scattered to the left and half to the right. One can either use an electrical network or a thermal capacitance argument to show that this results in ΔV being added to both of the scattered pulses:

$$\begin{aligned}
{}^{s}V_{L}(x) &= \rho\,{}^{i}V_{L}(x) + \tau\,{}^{i}V_{R}(x, y+1) + \Delta V \\
{}^{s}V_{R}(x) &= \tau\,{}^{i}V_{L}(x) + \rho\,{}^{i}V_{R}(x, y+1) + \Delta V.
\end{aligned} \tag{4.41}$$

In a situation where an input is time varying this will simply involve the adjustment of the amplitude of excitation at every time step.

(2) Constant temperature source
The implementation of a constant temperature source in bulk material is relatively complicated and will be omitted here. Constant temperature sources at a boundary are more common and easier to handle.

4.3.5 Boundaries

As in lossless TLM, the standard descriptions of short-circuit and open-circuit terminations can be used to model certain classes of physical boundaries in heat and matter simulations.

(1) An insulating boundary
Any heat approaching an insulating boundary is reflected back into the physical problem. This is the open-circuit ($\rho = 1$) condition and in both link-line and link-resistor formulations it is customary to place it at the interface between nodes. Thus a pulse travelling from a node during time $\Delta t/2$, encounters the boundary and arrives back at the node at the end of the time step.

(2) A symmetry boundary
As in lossless propagation, the computation of heat or matter profiles can sometimes be reduced by exploiting any symmetry in the problem, so that only half the problem need be simulated. The interface along the symmetry axis becomes an open-circuit boundary ($\rho = 1$).

(3) A perfect heat-sink boundary
This is covered by the definition of the short-circuit boundary of TLM, but some additional care is required. Once again, the boundary is placed at the interface between two nodes. There are however slight differences between the link-line and link-resistor formulations. In a link-line model the pulse is half-way along a transmission line when it sees a termination $Z_{T} = 0$ in front of it. The reflection coefficient is thus $\rho = -1$. In a link-resistor node the normal load impedance which a pulse sees as it reaches the end of the line is $R + R + Z$. One of these two resistors is associated with the node. The normal description

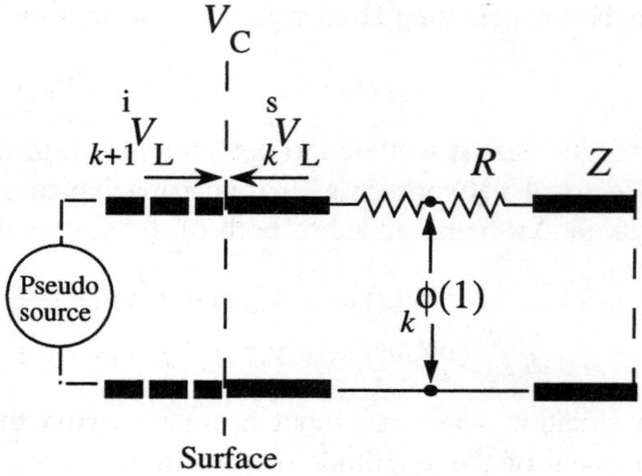

Fig. 4.30 The boundary node in a link-line formulation showing the 'ghost' source and line together with the incident and scattered pulses.

of a short-circuit condition in such cases is that the short is located immediately outside the node. Thus the line terminating impedance is $Z_T = R$ and the reflection coefficient from a short circuit is given by:

$$\rho = \frac{R - Z}{R + Z}. \tag{4.42}$$

(4) Constant temperature boundaries
In a link-line model the transmission line touches the boundary, whose value is held constant (V_C, as shown in Fig. 4.30).

In common with finite-difference treatments (Chapter 3) we will assume that there is a *ghost* node outside the boundary, and that this has a source and transmission line. This ensures that the value of the potential (V_C) at the surface, which is the summation of the pulse incident on node 1 at each new time step and the pulse scattered from node 1 at the previous time step is always constant.

$$_{k+1}{}^i V_L(1) + {}_k^s V_L(1) = V_C. \tag{4.43}$$

Since ${}_k^s V_L(1)$ is known at the present time-step, this equation can be used to calculate $_{k+1}{}^i V_L(1)$.

The situation with a link-resistor treatment is quite different since it is the resistor, not the transmission line, which touches the boundary. There are then two separate considerations:

• the input from the source which can now be situated *at* the boundary (Fig. 4.31(a)),

• the history of the pulse which is scattered from node 1 and which *approaches* the boundary (Fig. 4.31(b)).

Table 4.5 10 cm copper bar (part 4): connection to a 100 °C source

---

10 nodes, $\Delta t = 0.01$ s, $\rho = 0.98843$, $\tau = 0.01157$

At first time step we will have (100 °C)/2 injected as $_1{}^i V_L(1)$. This may not be accurate, but it is an acceptable first estimate and will work itself out after a few iterations.

$_1\phi(1) = 50$

$_1{}^s V_L(1) = 0.98843 * 50 + 0.01157 * 0 = 49.4215$

$_1{}^s V_R(1) = 0.01157 * 50 + 0.98843 * 0 = 0.5785$

$_2{}^i V_L(1) = 100 - 49.421 = 50.5785$

$_2{}^i V_L(2) = 0.5785$

$_2\phi(1) = 50.5785$, $_2\phi(2) = 0.5785$

$_2{}^s V_L(1) = 0.98843 * 50.5785 + 0.01157 * 0 = 49.993$

$_2{}^s V_R(1) = 0.01157 * 50.5785 + 0.98843 * 0 = 0.5851$

$_2{}^s V_L(2) = 0.98843 * 0.5785 + 0.01157 * 0 = 0.5718$

$_2{}^s V_R(2) = 0.01157 * 0.5785 + 0.98843 * 0 = 0.00669$

$_3{}^i V_L(1) = 100 - 49.993 = 50.007$

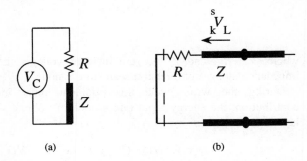

(a)   (b)

Fig. 4.31 (a) The network which is seen by the input at the boundary. (b) The situation which is seen by the pulse scattered towards the boundary.

The source (V_C) on the boundary sees a series connection of resistor and impedance, so that the standard potential divider formula gives the signal injected into the line. The pulse scattered towards the boundary sees a short circuit so that the total which is incident from the left at a new time step is the sum of these contributions:

$$_{k+1}{}^i V_L(1) = V_C \frac{Z}{R+Z} + {}_k{}^s V_L(1) \frac{R-Z}{R+Z}. \tag{4.44}$$

The running example is then concluded by implementing the connection of the copper bar to a 100 °C source and this is shown for several time-steps in Table 4.5.

4.3.6 Some examples of diffusion models using TLM

(1) Laser heating in rewritable CD ROMs

This application is outlined in the PhD thesis of Patel [4.9]. During the 'write' process in magneto-optic data recording the laser must raise the temperature of an active layer on the disk medium above a critical temperature in order to enable a change of logic state. The modeller will be interested in the laser conditions that ensure that this is done effectively at all points on the disk (the dwell time of the laser on a centre track will be much greater than on an outer track). The modeller may also be interested to know the maximum packing density that is possible while ensuring individual bits are distinguishable during the 'read' process.

Thermal diffusion after laser irradiation can be modelled using a three-dimensional TLM routine. The inclusion of stubs has been found to be very useful to account for the different material layers. This means that there are seven scattering and incidence directions per node. The nodal voltage is given by:

$$_k\phi(x, y) = \frac{\sum\limits_{m=1}^{7} \frac{2_k^i V_m}{R_m + Z_m}}{\sum\limits_{m=1}^{7} \frac{1}{(R_m + Z_m)}}$$

where $^i V_m$, R_m, and Z_m are the incident voltage, thermal resistance, and impedance along any direction seen from a node.

During the write cycle this equation must include an additional contribution: the energy input into a node of volume $V(x, y, z)$ in the active layer which is given by:

$$Q(x, y, z, t) = Q(x, y, t) \, \alpha \, e^{-\alpha(h-z)} V(x, y, z)$$

where α is the absorption coefficient of the active layer and h is the thickness. Also

$$Q(x, y, t) = \frac{P_0}{\pi v_x v_y} e^{-\left\{ \frac{(x - c_x \Delta t)^2}{v_x^2} + \frac{(x - c_y \Delta t)^2}{v_y^2} \right\}}$$

describes the incident energy during Δt due to irradiation by a laser of power P_0 with a geometry profile described by v_x and v_y as it passes over the disk with velocities c_x and c_y the laser velocities in the x and y direction (with respect to a moving sample).

Comparisons between experimental results and TLM models have highlighted the fact that the thermal conductivity in GdTbFe active films (1450 Å thick) is a strong function of temperature and is significantly less than that measured in bulk material.

(2) Modelling of viscous flow

The vertical flow of a viscous fluid from an opening was characterized by Trouton [4.10]. A decending fluid at any distance y from an orifice has an area $A(y)$ and a velocity $v(y)$ and a rate of elongation $dv(y)/dy$. The weight of material below y provides a tractive force such that:

$$\frac{F(y)}{A(y)} = \lambda \frac{dv(y)}{dy}$$

where λ is the coefficient of viscous traction and is given by $\lambda = 3\eta$ (where η is coefficient of viscosity). We can start to develop a TLM model of this process by considering an elemental volume of fluid of mass $\rho A(y) dy$ so that:

$$\frac{dF}{dy} + \rho A g = \rho A \frac{dv}{dt}.$$

If this substituted into the equation above we obtain

$$\frac{d^2 v}{dy^2} + \frac{\rho}{\lambda} g = \frac{\rho}{\lambda} \frac{dv}{dt} \qquad \left(\text{where } \frac{dv}{dt} = \frac{\partial v}{\partial t} + v \frac{\partial v}{\partial t} \right)$$

which is consistent with the Navier–Stokes equation. The problem can be described as diffusion of the effect of gravity causing relative motion within the body, the exact amount depending on the density and viscosity. This last equation could be rewritten as:

$$\frac{d^2 v}{dy^2} + \frac{\rho}{\lambda} \frac{d(gt)}{dt} = \frac{\rho}{\lambda} \frac{dv}{dt}.$$

The TLM equivalent for this would be:

$$\frac{d^2 \Phi}{dy^2} + 2R_d C_d \frac{d(\Phi')}{dt} = 2R_d C_d \frac{\rho}{\lambda} \frac{d\Phi}{dt}$$

where $\Phi' = g't$, and g' is constant. This is identical to any TLM diffusion model with a term $2R_d C_d \dfrac{d(\Phi')}{dt}$ which could be considered as driving the equivalent circuit.

In a practical implementation a contribution $g'\Delta t$ is made at each time step in the form of $g'\Delta t/2$ added to each incident pulse. Here Φ is the velocity, $\Delta x C_d$ is the elemental mass, and the elemental resistance is given by:

$$R = \frac{\Delta x}{2\lambda A(y)}.$$

The current is given by $I = C\dfrac{d\Phi}{dt}$ so that the net current between nodes is the viscous tractive force developed over a time period. For high values of λ it is valid to treat gravity as a current source of magnitude Cg.

Newton [4.11] has successfully used this approach to simulate the growth of fluid columns as part of research into the application of TLM to firing processes during the manufacture of vitreous china ware.

Readers might also like to consult papers by Patridge *et al.* [4.12] and Boucher and Kitsios [4.13] for further other TLM applications in mechanical engineering.

References

4.1 J. Vine, Impedance Newtorks, chapter in *Field analysis; experimental and computation* Vitkovitch (editor), Van Nostrand, London 1966.

4.2 P.B. Johns and R.L. Beurle, Numerical solution of 2-dimensional scattering problems using a transmission line matrix. *Proceedings of the IEE* **118** (1971) 1203–1208.

4.3 W.J.R. Hoefer, *The transmission line matrix method – theory and applications. IEEE Transactions on Microwave Theory and Techniques* **MTT-33** (1985) 882–893.

4.4 A. Amer, *The condensed node method and its applications in transmission in power systems.* PhD thesis, Nottingham University 1980.

4.5 P.B. Johns, *A symmetrical condensed node for the TLM method. IEEE Transactions on Microwave Theory and Techniques* **MTT-35** (1987) 370–377.

4.6 M.J. Buckingham and A. Tolstoy, An analytical solution for the benchmark problem I: the 'ideal' wedge. *Journal of the Acoustic Society of America* **87** (1990) 1511–1513.

4.7 D. Cheng, *Field and wave electromagnetics* (2nd edition), Addison Wesley 1989, pp. 449–452.

4.8 S.Y.R. Hui and C. Christopoulos, The modelling of networks with frequently changing topology whilst maintaining a constant system matrix. *International Journal of Numerical Modelling* **3** (1990) 11–21.

4.9 H. Patel. *Non-linear modelling of heat flow in magneto-optic multilayered media.* PhD thesis, Keele University, 1994.

4.10 F.T. Trouton, On the coefficient of viscous traction and its relation to that of viscosity. *Proceedings of the Royal Society,* **77** (1906) 426–440.

4.11 H. Newton, *TLM models of deformation and their application to vitreous china ware during firing.* PhD thesis, Hull University 1994.

4.12 C.J. Partridge, C. Christopoulos and P.B. Johns, Transmission line modelling of shaft system dynamics. *Proceedings of the Institue of Mechanical Engineers* **201** (1987) 271–278.

4.13 R.F. Boucher and E.E. Kitsios, Simulation of fluid network dynamics by transmission line modelling. Proc. I. Mech. E. **200** (1986) 21–29.

Examples, exercises, and projects

Most of the following problems have their origins within electronics. This is largely because the TLM method has grown out of electronics and uses many of the concepts from that discipline. These problems are largely developmental and give the reader the opportunity to develop familiarity with this relatively novel technique. The problems generally follow the sequence in the text and are intended to demonstrate the wider range of applicability. Most of the problems have several parts which can be tackled progressively.

4.1 A one-dimensional lossless propagation can be modelled by having a TLM mesh of 10 nodes where a point 3 nodes from one end is excited by means of a single impulse of magnitude 100. Investigate the effects of having the following terminations:

(a) open circuit at both ends

(b) short circuit at both ends

(c) a mixture of terminations.

It would be useful to choose a point (e.g. at the centre) and write the value at every iteration to a file. The data can then be analysed using a Fourier transform routine from Maple, Mathematica, or as outlined in Chapter 8.

The excitation position in the previous example (with two short-circuit boundaries) can be driven by a sinusoidal signal. We will define the time discretization as $T/20$, where T is the reciprocal of the driving frequency. In the first instance we could add a component $10 \sin[2\pi k(T/20)]$ to the voltage pulse scattered to the left at time step k. It would be worth comparing the Fourier response of this with a situation where an input $5 \sin[2\pi k(T/20)]$ was added to ^sV_L and ^sV_R at the kth time step.

In a further development of lossless one-dimensional scattering we can consider the effects of a nonlinear element in the circuit. Let us assume that this is placed half-way between nodes 6 and 7 as shown in Fig. 4.32.

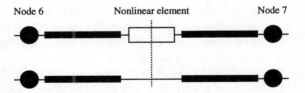

Node 6   Nonlinear element   Node 7

Fig. 4.32 Circuit for Problem 4.1.

This means that impulses coming from the nodes will experience and intermediate scattering event. The incident pulses at these two nodes will be:

$$_{k+1}{}^iV_R(6) = \rho_{NL}\ {}_k{}^sV_R(6) + \tau_{NL}\ {}_k{}^sV_L(7)$$

$$_{k+1}{}^iV_L(7) = \rho_{NL}\ {}_k{}^iV_L(7) + \tau_{NL}\ {}_k{}^sV_R(6).$$

Investigate the effect of this nonlinear element on the resonance properties of the network terminated in two short circuits if the impedance of the non linear element is given by:

$$Z_{\text{NL}} = \left[\frac{2}{\phi(6) + \phi(7)} \right]^2.$$

4.2 On p. 79 there is some code for a two-dimensional TLM algorithm. As it stands it is severely limited in its application. It was arranged for integer division and the excitation was designed to be divisible by 4 during 5 iterations. This should now be rearranged for real division with the option of having any value of excitation at any position in a larger array. It should also be adapted to include boundaries.

When this has been completed you should investigate the effects of impulse excitation and Gaussian excitation as shown in Fig. 4.33.

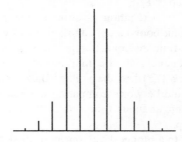

Fig. 4.33 Gaussian excitation.

The vertical axis represents amplitude of excitation. The horizontal axis would normally be discrete intervals of time (e.g. every time step or every m time steps) and these effects could be investigated in the time and frequency domain. Consider the problems that might arise in a two-dimensional model if the horizontal axis represented the spatial distribution of a single excitation in time.

A further adaptation of the two-dimensional code could involve the incorporation of stubs. The diagram shown in Fig. 4.34 might represent an optical fibre with its cladding. The waveguiding effect can be seen by driving the excitation point using a sinusoidal source. If a value of $\Delta x / \lambda = 0.1$ is chosen then show that an excitation $10 \sin(2\pi f t)$ is given by $10 \sin(0.2\pi k)$ where k is the iteration number. In view of the memory requirements of this problem it will be useful to make maximum use of symmetry properties. It will be noted that boundaries at the end walls have not been defined because any effects which they might introduce can be avoided if the algorithm is limited to less than 500 iterations.

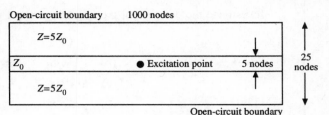

Fig. 4.34 Configuration for Problem 4.2.

Beaming effects of transmitters can be investigated using the two-dimensional routines which have already been developed. Start with a situation where two excitation points are set close together compared with the overall dimensions of the mesh. One of these is driven at $10 \sin(0.2\pi k)$ while the other is driven at $10 \sin(0.2\pi k + \pi)$. The effects of constructive and destructive interference can be observed in the time-domain response. The effects of different phase angles and additional excitation sources should also be investigated and the results can be compared with the theoretical predictions in standard texts such as Cheng [4.7].

4.3 The connect process in TLM can be done on the entire network using matrix transformations. Consider a matrix which has zero values except for one point where there is an incident pulse from the west:

$$
{}_{k}^{i}V_{\mathrm{W}} = \begin{bmatrix} 0 & 0 & 0 & 0 & 0 \\ 0 & 0 & 0 & 0 & 0 \\ 0 & 0 & 1 & 0 & 0 \\ 0 & 0 & 0 & 0 & 0 \\ 0 & 0 & 0 & 0 & 0 \end{bmatrix}
$$

and consider the shift matrices:

$$
S_1 = \begin{bmatrix} 0 & 1 & 0 & 0 & 0 \\ 0 & 0 & 1 & 0 & 0 \\ 0 & 0 & 0 & 1 & 0 \\ 0 & 0 & 0 & 0 & 1 \\ 0 & 0 & 0 & 0 & 0 \end{bmatrix}
$$

and

$$
S_2 = \begin{bmatrix} 0 & 0 & 0 & 0 & 0 \\ 1 & 0 & 0 & 0 & 0 \\ 0 & 1 & 0 & 0 & 0 \\ 0 & 0 & 1 & 0 & 0 \\ 0 & 0 & 0 & 1 & 0 \end{bmatrix}.
$$

Investigate the outcome of the following matrix products:

(a) $S_1 {}_k{}^i V_W$

(b) ${}_k{}^i V_W S_1$

(c) $S_2 {}_k{}^i V_W$

(d) ${}_k{}^i V_W S_2$.

These can incorporated in a scheme which would performs the TLM scatter/connect process over the entire network. With Matlab a very useful solution algorithm can be developed. If a symbolic package such as Maple is used then it is possible to obtain algebraic expressions for propagation at successive time steps.

4.4 The response of a circular membrane to an excitation has an analytical solution based in Bessel functions. Any attempt to form a numerical solution by normal methods will involve a stepped boundary and this can be confirmed by a simple TLM model.

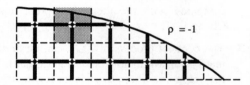

Fig. 4.35 Boundary definition for Problem 4.4.

The inclusion of more realistic boundaries requires a little ingenuity. If we take the shaded node in the section of a circular network which is shown in Fig. 4.35 we will see that one transmission line is of different length from the others. Therefore it does not have the same capacitance and therefore the impedance of this line will be different.

The impedance of a line of length a and a transit time Δt is $\Delta t/(C_d a)$. This is the same as:

$$\frac{\Delta t}{C_d \Delta x (a/\Delta x)} = Z_0 \frac{1}{(a/\Delta x)}$$

If the impedance of the boundary line at node (x, y) is $Z_N(x, y)$, then it is obvious that the reflection coefficient at this node will no longer be $-1/2$ but will be dependent upon the direction of incidence. Use the concepts embodied in Section 4.2 and surrounding text to develop an algorithm which will account for the behaviour of the nodes which are adjacent to the continuous boundary outlined here and use a discrete Fourier transform (DFT) to compare the results of an excitation with analytical predictions.

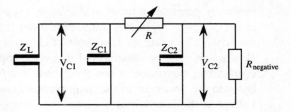

Fig. 4.36 Circuit for Problem 4.5.

4.5 Problem 3.3 can be re-examined using the TLM approach for lumped circuit simulations. The electrical circuit shown in Fig. 4.36 replaces the Chua circuit.

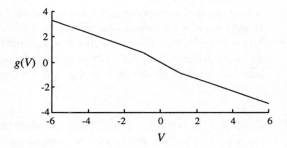

Fig. 4.37 Graph of conductance parameter.

The reciprocal of the conductance parameter, g which is redrawn in Fig. 4.37 represents the value of the negative resistance.

The objective in this exercise is to investigate the effects of Z_{C1} and $R_{negative}$ on the relative behaviour of V_{C1} and V_{C2}.

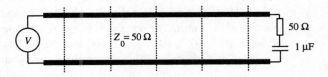

Fig. 4.38 Circuit for Problem 4.6.

4.6 Prepare an algorithm for a network of the form shown in the diagram of Fig. 4.38 which is intended to demonstrate the influence of a frequency dependent boundary.

The voltage generator is a random white noise source which delivers an input at every iteration which is given by $\sin(0.2\pi km)$ where m is a randomly generated number in the range $0 \leq m \leq 10$. The RC network at the other end represents a low-pass filter. It can be incorporated into the network by using the concepts of TLM modelling of lumped circuits. However it must be remembered that if the capacitor is replaced by an open-circuit transmission line then there is a half time step difference between the processes in the main problem and in the RC circuit. The voltage at the interface V_B becomes the driving force in the filter as in Fig. 4.16. The analysis (similar to eqns (4.22)–(4.25)) gives rise to a current. There is also a current in the network of transmission lines. These sum together to give a voltage at the boundary:

$$V_B = \frac{\frac{2^i V_C}{R+Z_C} + \frac{2^s V_R}{Z_0}}{\left[\frac{1}{R+Z_C} + \frac{1}{Z_0}\right]}$$

where Z_C is the impedance of the capacitor ($\Delta t/2C$) and $^s V_R$ is the voltage impulse scattered to the right at the node nearest to the boundary. The voltage V_B then determines the driving voltage of the RC network at the next iteration and the voltage scattered back into the line which becomes incident on the node at the next iteration:

$$^i V_R = V_B - {}^s V_R.$$

Once this is working it might be useful to use a single-frequency excitation source to determine the attenuation of the boundary as a function of frequency. Further extensions might involve a frequency selective boundary as in Fig. 4.16 and an extension to two-dimensional scattering problems.

4.7 The transition from lossless to lossy TLM can be observed by taking a one-dimensional network consisting of 20 nodes with short-circuit terminations at both ends and introducing a single impulse excitation moving from left to right. The algorithm uses $\rho = R/(R+Z)$ and $\tau = Z/(R+Z)$ where initially $R = 0$, $Z = 1$, and is run for several transits of the mesh. The process is re-run several times with the value of R set at 10^{-6}, 10^{-4}, 10^{-2}, 1, 10, 100 and the propagation of the pulse is observed at each stage. What conclusions can be drawn?

A lossy line can also be used to obtain a solution for the Laplace equation. Construct a one-dimensional lossy line with one boundary fixed at 1000 °C and the other set at 0 °C. Equation (4.43) can be used to launch impulses into the network which are then scattered using ρ and τ. The process is continued until the values of temperature at the nodes are within 0.001% of the analytical value (note that the two extreme nodes are only a distance $\Delta x/2$ from the boundaries). Investigate the number of iterations which are required in order to converge to this accuracy for a mesh of length of L nodes and a reflection coefficient ρ. There is an optimum ρ which gives fastest convergence for a given L and accuracy. Investigate how this changes with L and accuracy in the range $10^{-5}\%$ to 1%.

4.8 The manufacturer of a well-known brand of battery supplies a device which can be used to indicate status by connection between the two terminals. The idea is that this gives rise to dissipation in a nonlinear resistor and the resultant temperature rise causes a coating of liquid crystal to change colour. The resistance of a material is given by $R = \rho$ (length/area) where ρ is resistivity. This parameter and its coefficient temperature dependence can be obtained from reference sources such as *The Handbook of Chemistry and Physics*, published by CRC. During an interval Δt the energy dissipated is given by $I^2 R \Delta t$ and this appears as heat. The material in question is nichrome and its density and thermal data (specific heat and thermal conductivity) can be obtained from the same source as the electrical resistivity.

Nichrome film (0.01 mm thick)

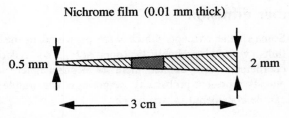

0.5 mm   2 mm

⟵ 3 cm ⟶

Fig 4.39 Battery status indicator.

Fig. 4.39 shows a typical battery status indicator which in thermal terms can be considered to be bounded by two constant temperature sources (the battery terminals; presuming perfect electrical contact) at 18 °C. The objective of this exercise is to discretize the problem and to produce a one-dimensional TLM heat-solver where the power dissipated is a function of the local electrical resistance, which in turn is a function of the local temperature at that time (obtained from $R(T) = R_0(1 + \alpha T)$). The overall objective is find a terminal current which ensures that the temperature within the shaded region at the centre of the resistor has reached 60 °C.

Rule-based models

5

This chapter represents a progression from the electromagnetic-based approach of the previous chapter. We will now consider the subject of cellular automata, lattice gas, and similar modelling techniques which represent a generalization of some of the concepts which were introduced earlier. The processes and models which we address here might be deemed to have an element of probability at the lowest level, but as a set of particles representing an ensemble of subparticles the behaviour is totally deterministic. Models based on completely random processes are reserved for the next chapter.

5.1 Number diffusion

Some of the concepts which were presented in the last chapter have direct links with a probabilistic approach. The reflection and transmission coefficients were defined there as electromagnetic entities, but they could equally represent probability weightings. The simple case $\rho = \tau$ for example can be considered in two ways:

- A quantity Q of electrical charge has an equal probability of being sent *on* (transmission) or sent *back* (reflected). A discussion on this discrete probability approach will be deferred to Section 5.3.

- A quantity Q of electrical charge is split equally so that half is reflected and half is sent onwards. This latter case is the subject of the next section.

5.1.1 Symmetrical number diffusion

The concept of equal probability of reflection and transmission allows us to use many of the ideas (and the code) that we presented in the last chapter. There is also a direct link back to the finite-difference ideas of Chapter 3, particularly the idea that every point is the average of the two points around it in a one-dimensional formulation:

$$_{k+1}V(x) = \frac{_kV(x+1) + _k V(x-1)}{2}. \tag{5.1}$$

Similarly, the value of a point in two dimensions is the average of the values of its four surrounding neighbours.

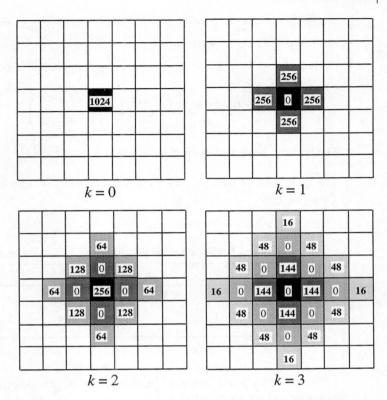

Fig. 5.1 Number diffusion during the first three time steps following an injection of 1024 at a source point.

Sticking with this idea, we can describe the diffusion of material injected at a point by saying that each particle of the material has equal probability of going in any of four directions in a Cartesian axis, so that one quarter of the original quantity will populate each of four positions at the next time instant. To that extent the process can be treated as deterministic. We can substitute a numerical value instead of a concentration of diffusing species and observe the process of number diffusion during several time-steps. An example of this is shown in Fig. 5.1 where the number 1024 is chosen for convenience (because the subdivision remains integral until after the fifth time step).

There are a number of very interesting points about this process and it is worth spending some time looking at the number sequences. If the origin, the point of injection at $t = 0$ is defined as $i = 0$, $j = 0$ then inspection of these numbers show that they follow what is called the *Bernoulli trial*:

(1) The population at position (i, j) at time k is zero for $|i| + |j| > k$.

(2) The population at position (i, j) at time k is zero for $k - i - j$ odd.

(3) For $|i| + |j| \leq k$ and $k - i - j$ even the population is given by:

$$\frac{(\text{initial input})}{4^k} \left(\begin{matrix} k \\ \frac{k-i+j}{2} \end{matrix} \right) \left(\begin{matrix} k \\ \frac{k-i-j}{2} \end{matrix} \right). \tag{5.2}$$

The terms in brackets are the binomial coefficients, i.e. $\left(\begin{matrix} k \\ m \end{matrix} \right) \equiv \dfrac{k!}{m!(k-m)!}$.

The application of the binomial coefficient can be easily checked by hand as shown in Table 5.1.

Table 5.1 The binomial coefficient $\left(\begin{smallmatrix} 4 \\ m \end{smallmatrix} \right)$

	$k = 4$				
m	0	1	2	3	4
$\left(\begin{smallmatrix} 4 \\ m \end{smallmatrix} \right)$	1	4	6	4	1

The binomial terms can also be determined by computer. Some simple recursive code is given below:

```c
#include "stdio.h"

int recur(int smaller,int larger);

void main(void)
{
int k, m;
int fac_m;
int fac_k_minus_m;
int fac_k;

printf("\nPlease enter integer values for k and m\n(k must be greater than m)\n? ");
scanf("%d%d",&k,&m);
fac_m=recur(0,m);
fac_k_minus_m=fac_m * recur(m,k-m);
fac_k=fac_k_minus_m * recur(k-m,k);
printf("\nResult is %f ",(float)fac_k/(fac_m * fac_k_minus_m));
}
```

```
int recur(int smaller, int larger)
{
if (larger>smaller)
        {
        return larger * recur (smaller, larger-1);
    }
else
        return 1;
}
```

Thus we could confirm that the value at $i = 2$, $j = 1$ at $k = 3$ after an initial input of 1024 is

$$_3V(2,1) = \frac{1024}{4^3}\left(\frac{3-2+1}{2}\right)\left(\frac{3-2-1}{2}\right) = 16\binom{3}{1}\binom{3}{0} = 48.$$

Similarly, the value at $i = 3$, $j = 1$ at $k = 4$ for the same input is

$$_4V(3,1) = \frac{1024}{4^4}\left(\frac{4-3+1}{2}\right)\left(\frac{4-3-1}{2}\right) = 4\binom{4}{1}\binom{4}{0} = 16.$$

As an example of the extension of these concepts we can consider a situation where a numerical value (1024 for convenience) representing a concentration of material is injected at location $(i = 0, j = 0)$ at time $k = 0$ and *also* at time $k = 1$. We would like to know the resulting effect at position $(i = 2, j = 3)$ at times $k = 7$ and $k = 8$. This can be expressed in terms of k as the super-position of two independent inputs:

$$\frac{1024}{4^k}\left(\frac{k}{k-2+3}\right)\left(\frac{k}{k-2-3}\right) + \frac{1024}{4^{k-1}}\left(\frac{k-1}{(k-1)-2+3}\right)\left(\frac{k-1}{(k-1)-2-3}\right).$$

$(k = 0 \text{ input})$ $\qquad\qquad\qquad\qquad$ $(k = 1 \text{ input})$

The first condition $(|i| + |j| \leq k)$ is satisfied for both $k = 7$ and $k = 8$. However at $k = 7$ the second condition $(k - i - j$ even) is only satisfied by the term due to the $(k = 0)$ input, since $(7 - 1) - 2 - 3$ is odd. On the other hand, at $k = 8$ the second condition is not fulfilled for the $(k = 0)$ input but is satisfied for the $(k = 1)$ input. The reader might like to calculate these terms to see that the value at $i = 2$, $j = 3$ only changes at every second time step. The effects at $i = 2$, $j = 3$ at times $k = 7$ and $k = 8$ due to inputs at $k = 0$ and $k = 2$ could also be investigated.

We can progress from here to consider a piece of material which is absorbing electrical energy at *every* instant in time (constant heat input) so that there is an addition to the temperature of T_{in} at $k = 0, 1, 2, 3 \ldots$

The temperature at position (i, j) and time k is then given by:

$$
_k T(i,j) = \frac{T_{in}}{4^k} \binom{k}{\frac{k - i + j}{2}} \binom{k}{\frac{k - i - j}{2}} +
$$

$$
\frac{T_{in}}{4^{k-1}} \binom{k - 1}{\frac{(k - 1) - i + j}{2}} \binom{(k - 1)}{\frac{(k - 1) - i - j}{2}} +
$$

$$
\frac{T_{in}}{4^{k-2}} \binom{k - 2}{\frac{(k - 2) - i + j}{2}} \binom{(k - 2)}{\frac{(k - 2) - i - j}{2}} +
$$

$$
\frac{T_{in}}{4^{k-3}} \binom{k - 3}{\frac{(k - 3) - i + j}{2}} \binom{(k - 3)}{\frac{(k - 4) - i - j}{2}} + \quad \ldots
$$

Using the idea of heat flow as an anchor we can extend the basic concept to estimate the temperature at position (i, j) and time (k) due to *any* inputs at other positions (i', j') and times (k'). The most general response at (i, j) and time k due to a single injection of magnitude $T_{in}(i', j', k')$ is given by:

$$
_k T(i,j) = \frac{1}{4^{(k-k')}} \binom{k - k'}{\frac{(k - k') - (i - i') + (j - j')}{2}}
$$

$$
\binom{k - k'}{\frac{(k - k') - (i - i') - (j - j')}{2}} T_{in}(i', j', k')
\tag{5.3}
$$

so long as $| i - i' | + | j - j' | \leq (k - k')$ and $(k - k') - (i - i') - (j - j')$ is even. Otherwise it is zero.

A temperature input of 512 at location $i' = 1, j' = 1$ at time $k' = 1$ will have an influence at $i = 2$, $j = 2$ at time $k = 3$ given by:

$$
\frac{1}{4^{(3-1)}} \binom{3 - 1}{\frac{(3 - 1) - (1 - 2) + (2 - 1)}{2}} \binom{3 - 1}{\frac{(3 - 1) - (1 - 2) - (2 - 1)}{2}} 512
$$

$$
= \frac{1}{4^2} \binom{2}{2} \binom{2}{1} 512 = 64.
$$

Because of the very close similarity of these concepts to Green's functions [5.1] we define a *discrete Green's function* as:

$$_{(k-k')}G(i,j;i',j') = \frac{1}{4^{(k-k')}} \left(\frac{k-k'}{(k-k')-(i-i')+(j-j')} \right)$$
$$\left(\frac{k-k'}{(k-k')-(i-i')-(j-j')} \right).$$

In a problem involving multiple excitations in space and time each input is regarded as propagating independently and the resultant value at any position and subsequent time is the sum of the contributions from the individual discrete Green's functions weighted by the corresponding excitation strengths. Thus, for a series of temperature inputs, $_{k'}T^s(i',j')$ in an infinite two-dimensional space we have:

$$_kT(i,j) = \sum_{k'=0}^{k} \sum_{j'=-\infty}^{+\infty} \sum_{i'=-\infty}^{+\infty} {}_{k-k'}G(i,j;i',j') \, _{k'}T^s(i',j'). \tag{5.4}$$

We will return again briefly to this topic in Chapter 9 when we will describe Green's functions in terms of integral equations.

5.1.2 Probabilistic potential theory

This process of number diffusion has been developed by Bevensee [5.2] under the title, *Probabilistic potential theory* to estimate Laplacian fields. It is left as an exercise to the reader to take the problem shown in Fig. 3.2 and treat all the node points at the boundaries as sources with either T^s(boundary) $= 0$ or 100 V. Equation (5.4) can be applied in order to estimate the potential of an individual node at any time. The problem as posed is in fact dynamic. At $k = 0$ the potential within the enclosed space is zero but develops as information about the boundaries diffuses throughout the problem space. Ultimately, it reaches a steady state and it is even be possible, although not easy, to derive an algebraic expression for the asymptotic value of potential.

5.1.3 Asymmetric number diffusion

The previous sections were concerned with a situation which was equivalent to the condition in transmission line matrix (TLM) modelling where $\rho = \tau$. Again, there is no absolute requirement that this should be the case and we will discuss here some consequences of removing that restriction. For simplicity, we will consider a one-dimensional situation where ρ and τ are the probabilities of reflection and transmission. Note that these are defined with respect to the direction of motion of an individual particle (or charge) and not with respect to any coordinate axis. This can be visualized in a scatter diagram

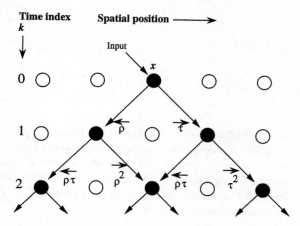

Fig. 5.2 Asymmetric scatter diagram for an initial input from the left.

such as that shown for an initial input from the left in Fig. 5.2. At time $k = 0$ the input is scattered ρ and τ. These are then independent entities which undergo scattering at the next iteration. Thus τ moving to the right is scattered to yield $\rho\tau$ moving to the left and $\tau\tau$ moving to the right. Similarly ρ moving to the left becomes $\rho\tau$ continuing leftwards and ρ^2 which is now moving to the right.

An inspection of this diagram shows some very interesting properties. If it is extended further it will be seen that the population scatter diagram is as in a Pascal triangle. In spite of this binomial process, it is not easy to predict the magnitudes of the contributors at any point.

Much information on this can be obtained using simple shift operators. As shown in the diagram above each contributor requires two description parameters, magnitude and direction of motion. These can be quantified by a reflection operator $\hat{\rho}$ and a transmission operator $\hat{\tau}$.

The operator $\hat{\rho}$ has two effects:

- It alters the pulse magnitude by a factor ρ.

- It reverses the direction of movement of a pulse, and shifts it by one position.

Thus, if an input of unit magnitude, arriving at position x from the left is designated as (x), then

$$\hat{\rho}(x) = \rho(x - 1). \tag{5.5}$$

If $\hat{\rho}$ acts on this then

$$\hat{\rho}\rho(x - 1) = \rho^2(x - 1 + 1) \text{ i.e.} \rho^2(x).$$

This is the value at the origin, which is exactly what is observed in the diagram.

The transmission operator $\hat{\tau}$ has two effects:

- It alters the pulse magnitude by a factor τ.

- It maintains the direction of movement of a pulse, and shifts it by one position.

For the pulse incident at location x from the left we can write

$$\hat{\tau}(x) = \tau(x+1). \tag{5.6}$$

These operators are associative, but not commutative. It is quite clear from an inspection of Fig. 5.2 that the sequence of operations $\hat{\rho}\hat{\tau}$ and $\hat{\tau}\hat{\rho}$ have identical effects on the magnitude of a pulse, but they result in pulses with quite different final locations.

If we now start the scatter process at the top of the triangle in Fig. 5.2 then it is obvious that the pulse from the left is subject to both the transmission *and* reflection shift operators. This can be summarized as:

$$(\hat{\rho} + \hat{\tau})(x) = \rho(x-1) + \tau(x+1).$$

At the next iteration time step ($\hat{\rho} + \hat{\tau}$) acts upon the previous result

$$
\begin{aligned}
(\hat{\rho} + \hat{\tau})(\rho(x-1) + \tau(x+1)) &= \rho^2(x-1+1) + \rho\tau(x+1-1) \\
&\quad + \tau\rho(x-1-1) + \tau^2(x+1+1) \\
&= \tau\rho(x-2) + \rho^2(x) + \rho\tau(x) + \tau^2(x+2).
\end{aligned}
$$

The scattering of two pulses has resulted in four pulses which at the instant of inspection are located as in a Pascal triangle: there is one pulse at positions $(x-2)$ and $(x+2)$, while there are two pulses at position x. At the next time instant there will be a total of 8 pulses (one each at $(x+3)$ and $(x-3)$, and three each at $(x+1)$ and $(x-1)$).

Table 5.2 The first three scattering events

$k = 1$	$k = 2$	$k = 3$
ρ	$\rho\rho$	$\rho\rho\rho$
τ	$\rho\tau$	$\rho\rho\tau$
	$\tau\rho$	$\rho\tau\rho$
	$\tau\tau$	$\rho\tau\tau$
		$\tau\rho\rho$
		$\tau\rho\tau$
		$\tau\tau\rho$
		$\tau\tau\tau$

It will of course be observed that the sequence develops as shown in Table 5.2. It is not surprising that this is closely related to the development of the binary number sequence and indeed if one uses a mapping between 0 and ρ and 1 and τ then some interesting number spatial relationships can be observed in the decimal equivalents of the operator sequences at different positions.

The scattering process can be modelled in software using a variety of approaches. The Mathematica code shown below treats finite space as the superposition of a pair of lists and uses string handling operations to acomplish the objective.

```
Rlist:= {0,0,0,0,0,0,0,0,τ,0,0,0,0,0,0,0}
Llist:= {0,0,0,0,0,0,0,ρ,0,0,0,0,0,0,0,0}
Temp:= (Rrtemp=Totate Right [Rlist,1]τ; Rltemp=Rotate Right [Llist,1]ρ;
    Lrtemp=Totate Left [Rlist,1]ρ; Lltemp=Rotate Left [Llist,1]τ)
Sums:= (Sumleft=Lrtemp + Lltemp; Sumright=Rrtemp + Rltemp)
Output (Results=Sumleft+Sumright;
    Do[Print{Expand[Results[ [i]]]},{i,Length[Rlist]}])
Swaps:=(Rlist=Sumright; Llist=Sumleft)
Iterate:= (Temp; Sums; Swaps; Output)
```

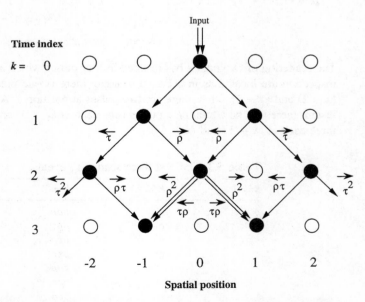

Fig. 5.3 A symmetric input leading to two independent asymmetric random walk processes.

So far, this scatter operator treatment has only considered a pulse which was initially incident from the left. The treatment for a pulse, initially incident from the right is the mirror image with $\hat{\rho}(x) = \rho(x+1)$ and $\hat{\tau}(x) = \tau(x-1)$. Otherwise the processes are identical.

These ideas have been presented in order to demonstrate the importance of a clear understanding of initial conditions. The two scattering processes could be described by the diagram in Fig. 5.2, or its mirror image, so long as the initial direction was known. We could return at this point to the previous chapter and use electrical network theory to deduce the nature of the initial input, as we discussed in Section 4.3.4 (Fig. 4.22). Regardless of the ρ/τ asymmetry, an input will have an equal probability of going left or right. Thus an input of magnitude 2 at $x = 0$ and $t = 0$ will result in a unit input from the left at $x = 1$, $t = 1$ and from the right at $x = -1$, $t = 1$. This is shown in Fig. 5.3 and if it is calculated for a single shot input, it will give results which are identical with analytical results.

Example 5.1 Minority carrier recombination in a photodiode

Photodiodes have a long history as technologically significant electronic components (photographic exposure meters, burglar alarms, etc.). They are once again assuming prominence with the advent of high-speed optical fibre communications. Light falling on the diode causes a local disturbance in the distribution of charge (changes the logic state to '1'). The charge which is introduced as a result of illumination (see Fig. 5.4) will diffuse in a manner which can be well described by the scattering dynamics of the previous section. However there is an additional effect. The diode will attempt to restore the situation to what it was before the illumination and this will result in recombination: a decay of the excess charge as time progresses. Just as in a first-order chemical reaction, the amount of remaining charge drops off exponentially with time and has a characteristic constant called the carrier lifetime (designated as τ_p for holes or τ_n for electrons).

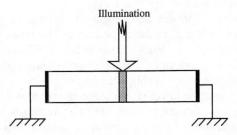

Fig. 5.4 An unbiased photodiode under illumination. The excess charge generated by the light will decay as it diffuses.

> Recombination can be included in the scatter process. For example, if we consider an assembly of holes $Q_p(x)$ coming in from the left of position x at time k, then the history at the next time step can be described using shift operators with a modification which takes account of the decay which occurs during the time interval Δt between scattering at node x and arrrival at adjacent nodes:
>
> $$(\hat{\rho} + \hat{\tau})Q_p(x) = \rho Q_p(x-1)\,e^{-\Delta t/\tau_p} + \tau Q_p(x+1)\,e^{-\Delta t/\tau_p}.$$
>
> When this is applied to the analysis, it provides results which are very close to experimental observations.

Diffusion with drift

It is now time to talk about another form of asymmetry. In this case it refers to a probability bias with respect to a fixed external axis. Let us start with a natural process that had reflection and transmission probabilities ρ and τ with respect to the individual direction of motion. Now we will assume that there is some external factor which biases ρ and τ by an amount α so that there is a net motion towards the right. In terms of the last example this might be due to a negative electrical potential on the right-hand terminal of the diode pulling charge from left to right. This bias would manifest itself in the scattering equations as:

$$_k^sQ_L(x) = (\rho - \alpha)_k{}^iQ_L(x) + (\tau - \alpha)_k{}^iQ_R(x)$$

$$_k^sQ_R(x) = (\tau + \alpha)_k{}^iQ_L(x) + (\rho + \alpha)_k{}^iQ_R(x).$$

$$(5.7)$$

In order to avoid physical oddities (such as reversal of the sign of the charge) the value of α must be such as to ensure that $(\rho - \alpha)$ and/or $(\tau - \alpha)$ are always greater than or equal to zero and likewise $(\rho + \alpha)$ and/or $(\tau + \alpha)$ are always less than or equal to unity.

These concepts can immediately be applied to model the chromatographic process which lead to the separation of chemicals by diffusion and drift in a column of solvent. Each chemical has a different adsorption isotherm (see Example 3.5) and thus different effective values of ρ and τ. The bias factor α affects both equally. If this were part of a police check on alcohol content the sample would be injected into a chromatography column along with a known quantity of reference material. The two materials would traverse the column and the relative times of arrival of sample and reference would confirm whether the sample was ethyl alcohol. The areas under each curve would give the relative concentrations.

The concept of diffusion–drift modelling of the chromatographic process can be extended to two dimensions where the flow of solvent carries the sample in a vertical direction over a sheet of adsorbant material. The scattering parameters in eqn (5.7) will have a bias α_v which depends on the flow rate. The simultaneous application of electrical potential across the sheet will cause certain chemicals such as amino acids and proteins to drift in a horizontal direction (electrophoresis). In this case the scatttering parameters will have a bias α_h. The practical application of this idea can lead to a clear identification of different amino and is the basis of genetic 'fingerprinting'.

In terms of semiconductor charge transport a simple model which contains diffusion, drift, and recombination applied to the structure in Fig. 5.4 yields results which are in very close agreement with the measurements of an experiment due to Haynes and Shockley [5.3] which confirmed many of the concepts of charge transport in semiconductors and provided actual values for the minority carrier lifetime.

Example 5.2 The dispersion of smoke from a factory chimney

Some of the concepts in Sections 5.1.1 and 5.1.3 can be brought together in a problem which was examined analytically by Crank et al. [5.4]. Smoke issuing from a factory chimney will disperse diffusely in three dimensions (assuming that buoyancy effects can be ignored). If however there is a strong wind then this will lead to drift/diffusion in one direction (e.g. x), whilst the processes in the y and z directions are purely diffusive. This is shown in Fig. 5.5. For the purposes of simplicity let us assume that the probability of transition in any direction is 1/6 so that $\rho + 5\tau = 1$. Thus the scattering process is biased by a factor α in the x axis only. The total process can then be described by the scatter matrix equation:

$$
\begin{bmatrix} {}^sQ_1 \\ {}^sQ_2 \\ {}^sQ_3 \\ {}^sQ_4 \\ {}^sQ_5 \\ {}^sQ_6 \end{bmatrix}_k = \frac{1}{6} \begin{bmatrix} 1 & 1 & 1 & 1 & 1 & 1 \\ 1 & 1-6\alpha & 1 & 1-6\alpha & 1 & 1 \\ 1 & 1 & 1 & 1 & 1 & 1 \\ 1 & 1+6\alpha & 1 & 1+6\alpha & 1 & 1 \\ 1 & 1 & 1 & 1 & 1 & 1 \\ 1 & 1 & 1 & 1 & 1 & 1 \end{bmatrix} \begin{bmatrix} {}^iQ_1 \\ {}^iQ_2 \\ {}^iQ_3 \\ {}^iQ_4 \\ {}^iQ_5 \\ {}^iQ_6 \end{bmatrix}_k
$$

where the directions are as defined in the inset in Fig. 5.5.

The subsequent connection process in three dimensions can then be expressed as

$$_{k+1}{}^iV_1(x,y,z) = {}_k{}^sV_3(x,y-1,z)$$

$$_{k+1}{}^iV_2(x,y,z) = {}_k{}^sV_4(x-1,y,z)$$

$$_{k+1}{}^iV_3(x,y,z) = {}_k{}^sV_1(x,y+1,z)$$

$$_{k+1}{}^iV_4(x,y,z) = {}_k{}^sV_2(x+1,y,z)$$

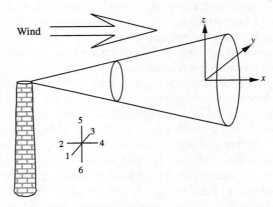

Fig. 5.5 Dispersion of smoke from a factory chimney under windy conditions. The scattering directions are shown as an insert.

$$_{k+1}{}^iV_5(x,y,z) = {}_k{}^sV_6(x,y,z+1)$$
$$_{k+1}{}^iV_6(x,y,z) = {}_k{}^sV_5(x,y,z-1).$$

This process can be repeated for as many iterations as required and α can be varied to simulate different wind speeds. This is as far as we will take this example, since it has served to indicate the approach and further development is beyond the scope of the illustration. Enthusiastic readers might wish to make suitable choices for Δx, Δy, Δz, and Δt to decide on the minimum computational space and to relate wind speed to the bias factor α.

Temporary storage at a node

In Section 4.2 the concept of stubs was introduced as a means of altering the local permittivity. There is another way of looking at this: a stub is a capacitance which holds charge during one time step. It is a temporary storage or delay element and this can also be interpreted in probabilistic terms.

We will start by considering a one-dimensional problem with the normal reflection and transmission probabilities. We can now introduce an additional parameter σ which represents the probability of storage at a node during one time step. The sum of these probabilities $(\rho + \tau + \sigma)$ is unity. Thus, instead of modelling adsorption (as in chromatography) by altering ρ and τ to take account of an effective diffusion constant, we could represent adsorption as a probability of storage through σ. These ideas could have interesting applications in the modelling of trapping processes. They can also arise in a 'natural' way in cellular automaton modelling which is discussed in the next section.

5.2 Cellular automata models

Introduction

The last section dealt with number diffusion (symmetrical and asymmetrical). Clear links to rules based on electromagnetics were indicated. This section goes a stage further to cover a class of models in which a population on a mesh is subject to the repeated application of a set of apparently *ad hoc* rules which nevertheless yields an ordered outcome.

Over the past twenty-five years, scientists have been studying a class of mathematical models known as *cellular automata* (CA). In these models, a small set of values is defined, including zero. Space consists of a regular grid of cells, each of which contains a value from this set. Time consists of a sequence of discrete intervals. At each interval a 'snapshot' of the grid shows the distribution of values — the *population* — at that time. The population is transformed over time by the application of a small number of extremely simple rules.

It turns out that such models can reproduce very complex phenomena. The application of the rules can lead to interesting patterns and behaviour even in the simplest case. The extension to slightly more complicated rule sets such as the famous 'game of life' leads to some fascinating time and space dependent patterns. The consequences of many of these ideas are raising some very fundamental questions for theoreticians.

This section starts by considering one of the simplest cellular automata of all. This leads us on to an investigation of the game of life. C code is provided here, so that readers can experiment with their own initial patterns. Next we look briefly at a simplified model for a computer known as the Turing machine. We then extend the discussion to consider (in outline) the application of CA to some practical examples. We close with a few general comments on the properties of cellular automata which have relevance in models of complex systems, and which may indicate the future direction of CA research.

5.2.1 Cellular automata in one dimension

In infinite one-dimensional space the grid of values is treated as indefinitely long, (finite CAs are also possible, but there must be a clear set of rules to determine behaviour at the boundaries). We start by describing the values, initial state and rules for a very simple infinite model. Values on the line are constrained to be either one or zero ('on' or 'off'). Initially (at time zero), the value of a single point is set to 1, all other values are set to 0. For convenience, we will suppose that the position with value 1 is $x = 0$.

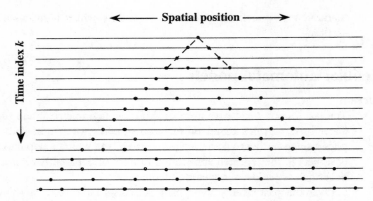

Fig. 5.6 The result of applying eqn (5.8) over many time steps. The black spots represent the odd value at every point on the mesh. It can be seen that each triangle is composed of repeats of a subunit (shown in outline).

At each time step k, every point on the line updates its state according to the following rules:

$$_{k+1}C(x) = 1 \quad \text{if } _kC(x-1) + _kC(x+1) \text{ is odd}$$

$$(5.8)$$

$$_{k+1}C(x) = 0 \quad \text{if } _kC(x-1) + _kC(x+1) \text{ is even.}$$

This means that those cells with an odd number of live neighbours at time k turn themselves on (become 1) at time $k+1$, while those with an even number of live neighbours turn themselves off (become 0). Mathematically, what we are seeing here is the result of modulo-two arithmetic. Modulo two is a binary operator which yields one if the sum of its operands is odd, zero otherwise.

The easiest way to visualize how this cellular automaton evolves is to draw its successive states below one another as in Fig. 5.6. In this computer printout each row shows one further step in time. Note particularly that the pattern at the apex of the triangle repeats itself again and again as time progresses. This self-reproduction of patterns (not necessarily the initial pattern) is a feature of many CAs and is also the basis of fractal behaviour.

5.2.2 General considerations

Certain other features of CAs deserve attention. While generally applicable, they are easiest to visualize in one dimension, and are therefore presented at this point.

1. Different update rules give different results. In the one-dimensional case we considered the node itself and its two nearest neighbours (3 sites in all), together with the set of possible states they could assume. In our example

there were two states, 1 and 0, and 8 possible conditions: of the three sites involved in any transformation. These could be represented by the binary numbers 000 to 111. There are many possible rules. Of these, some are not very useful and are described as *illegal* [5.5]. Each *legal* rule (there are 32 of these) has slightly different general characteristics and statistics from the others. With more states the number of possible rules increases dramatically.

2. Although the patterns that develop from a given starting configuration are predetermined for any rule set, it is almost always impossible to envisage future patterns without carrying out the actual regenerative processes contained in the rules. Sometimes the 'society' eventually dies out; possibly after several apparently 'healthy' generations. Other starting patterns converge to stability (shown later in Fig. 5.7(c) or oscillate forever (Fig. 5.7(d)).

3. Mathematical models of linear systems tend to approach a state of equilibrium over time. Nonlinear systems generally settle into equilibrium or oscillating states. Many of the same characteristics are observable in CAs. Random binary numbers introduced as initial conditions often produce patterns that become gradually more organized and symmetrical, resembling their mathematical counterparts.

4. Perturbations can be introduced into any system by inserting values (noise) at different positions and times after the start time. Systems will attempt to absorb such interference, but if the noise is sufficiently extensive and random then the system will never become stable. Such chaotic behaviour is also seen in finite CAs where the boundaries are nondeterministic.

5. In a later section (5.2.4) on the statistics of CAs we will consider analogies with the second law of thermodynamics ('the entropy of the universe is rising'). In this context the question of reversibility of processes is important. Wolfram [5.4] defines an irreversible CA as one which leads to configurations which cannot be reached by evolution from any other configuration and which can only appear as initial sites. It has been shown that while certain rule sets produce CAs that are reversible, this is not always the case. The game of life is just such an example of irreversibility.

5.2.3 Two-dimensional CAs

In two dimensions it is possible to distinguish between orthogonal and diagonal neighbours. This leads to types of CAs with different numbers of sites and significantly different numbers of possible rules. Nine-site CAs (acknowledging both type of neighbour) have 271 rules of which 259 are legal.

The CA known as *Life* was an early example of the 9-state binary rule set devised as an entertainment by the mathematician John Horton Conway in

1970. It was first developed as a rectangular solitaire-type board game involving a set of counters similar to those used in the game Go. The spread of computers automated the process of rule application and allowed investigation of long-term behaviour and this was exploited by Martin Gardner in several articles in Scientific American [5.6, 5.7].

The basic idea is to start with a simple two-dimensional configuration (population pattern) of cells (organisms) on a grid and observe how the pattern of *live* cells changes as a particular rule set is applied. Conway was interested in mimicking a form of *genetic* behaviour involving births, lifespans and deaths of organisms over a period. His rule set was formulated, after much experimentation, to fulfil three conditions:

1. There should be no initial pattern for which there is a simple proof that the population can grow without limit.

2. There should be initial patterns that apparently do grow without limit.

3. There should be simple initial patterns that grow and change for a considerable period of time before coming to an end in three possible ways:

(a) dying away completely;

(b) converging to a stable configuration;

(c) entering an oscillation cycle of two or more periods.

Conway's genetic laws are straightforward. The value of any point in a two-dimensional grid is either 1 (live) or 0 (dead). At time zero a pattern of live cells is introduced onto the grid by setting certain values to 1. All other values are set to 0. At each time step k, every point on the grid updates its state (depending on the states of the eight neighbouring cells) according to the following rules:

1. *Survivals*: Every 'live' cell with two or three 'live' neighbours survives.

2. *Deaths*: (i) Every 'live' cell with four or more 'live' neighbours dies from overcrowding. (ii) Every 'live' cell with less than two 'live' neighbours dies from isolation.

3. *Births*: Every 'dead' cell adjacent to exactly three 'live' neighbours comes to life.

This is accomplished in practice by maintaining two arrays, one showing the population pattern of the current generation, the other receiving the values for the new generation (the next time step) as they are calculated.

Given below is an example of a game of life program written in Turbo C. We start here with a 21 by 21 mesh on which we have defined a population. This population is updated by the application of the Conway rules every time the space bar is pressed.

```c
#include "stdio.h"
#include "conio.h"

void main (void)
{
int oldmesh[21][21], newmesh[21][21];

int i,j;                              /* x and y counters */
int sum;                             /* sum of population of surrounding grid points */
int xbefore,xafter,yabove,ybelow;    /* surrounding grid points */
char go = ' ';                       /* procreate until go is set to non-space character */

printf("\n\n Life example program\n\n");
printf("Instructions: use space bar for another generation, any other key to stop\n\n");

/ *
* set everything to zero (unpopulated)
* /

for(i=1;i<21;i++)  for(j=1;j<21;j++)  oldmesh[i][j]=newmesh[i][j]=0;

/ *
* set a row of three centrally placed cells to one (populated)
* the interested user can easily change this pattern
* /

i=10;  for(j=8;j<11;j++)  oldmesh[i][j]=1;

while(go==' ')
   {
   / *
   * calculate the sum of surrounding population for each cell in the model and
   * use Conway's rules to determine whether cell will be populated in next generation
   * /

   for(i=1;i<21;i++)
      {
      for(j=1;j<21;j++)
         {
         / *
```

```
* this model 'wraps around' so that the column furthest left
* is adjacent to the column furthest right, and the
* lowermost row is adjacent to the uppermost row
* /
xbefore=(i==1)?20:i-1;
xafter=(i==2)?1:i+1;
yabove=  (j==1)?20:j-1;
ybelow=  (j==20)?1:j+1;

/ *
* calculate sum
* /
sum=oldmesh[xbefore][yabove]+oldmesh[xbefore][j]+
oldmesh[xbefore][ybelow]+oldmesh[i][yabove]+

oldmesh[i][ybelow]+oldmesh[xafter][yabove]+
oldmesh[xafter][j]+oldmesh[xafter][ybelow];

/ *
* apply rules
* /
switch (sum)
    {
    case 3: newmesh[i][j]=1; break;
    case 2: if(oldmesh[i][j]) newmesh[i][j]=1; break;
    default: newmesh[i][j]=0;
    }
  }
}

/ *
* display mesh and replace old mesh with new  mesh
* /
for(i=1;i<21;i++)
    {
    printf("\n");
    for(j=1;j<21;j++)
    {
        printf("%2d",newmesh[i][j]);
        oldmesh[i][j]=newmesh[i][j];
    }
  }
printf("\n\n");
go=getch();
  }

}
```

By editing this program you can create your own populations and watch how they develop. Some examples which could be tried are shown in Fig. 5.7.

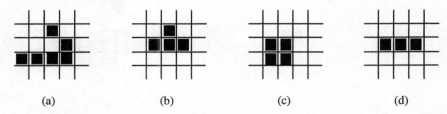

(a)   (b)   (c)   (d)

Fig. 5.7 Some elementary life configurations.

Example 5.3 Wolfram's wires — a digital circuit simulator

This example is concerned with building a set of components which can be put together in various formations to simulate digital electronic circuits and ultimately entire computer architectures. This, and Life representations of logic gates, were proposed in 1983 by Stephen Wolfram [5.5]. Practical implementations have been discussed by the computer journalist Wilf Hey in two articles [5.8, 5.9].

A circuit consists of a set of adjacent nodes on a two-dimensional mesh. Such a configuration is called a wire. The passage of electricity along a wire is represented by leapfrogging head/tail pairs. The rule set at each time step is as follows:

(1) a head becomes a tail;

(2) a tail becomes a wire;

(3) a wire becomes a head if it borders one or two heads, otherwise it remains a wire;

(4) All other cells in the mesh are unaffected.

Before these rules can be applied in practical circuits, we must develop three basic components:

1. *A diode* This is shown in Fig. 5.8 which also shows the sequence of current flow in both directions. It can be seen that this configuration

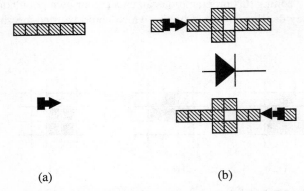

(a)   (b)

Fig. 5.8 (a) A wire and a head–tail pair indicating the direction of current flow. (b) A diode with its *Wires* equivalent showing the concept of forward and reverse bias. The application of the rules will cause the annihilation of the pair in the reverse-biased diode.

ensures unidirectional flow (it is useful to insert diodes in any stretch of wire to inhibit back flow of current).

2. *An oscillator* An input into a circuit can always be achieved by placing a head/tail pair at one end of a short length of wire attached to the circuit. For repetitive input signals it is useful to have an oscillator such as the one shown in Fig. 5.9(a). This is a loop of wire attached to a diode which emits a head/tail pair through the diode once per cycle. The length of the wire loop determines the signal repetition rate.

3. *A sink* This component, shown in Fig. 5.9(b). effectively removes head/tail pairs from a circuit. It is the *Wires* equivalent of electrical ground. It comprises a loop beyond a diode which ensures that any signal entering the loop does not return. An additional diode is placed inside the loop to forces one-way flow.

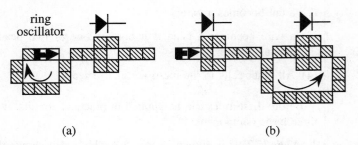

ring oscillator

(a)   (b)

Fig. 5.9 (a) An oscillator. (b) A sink. The diagrams also show the diodes and the direction of current flow.

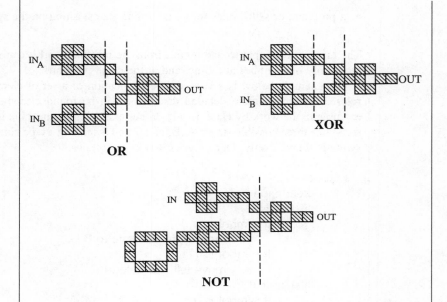

Fig. 5.10 OR, XOR (exclusive OR), and NOT gates. The oscillator in the NOT gate is designed to produce a high input to the XOR gate at all times so that the output will always be the opposite of the input.

Figure 5.10 shows the required configurations for an OR gate, an XOR gate, and a NOT gate. With these three gates it should be possible, (though perhaps unrealistic in practice) to construct any logic function.

Example 5.4 The Turing machine
Smith [5.10] showed that Life is capable of providing a simple model of a rudimentary computer, the Turing machine. This was named after the English mathematician Alan Turing, who showed that any computation that could be carried out by any conceivable computer could also be carried out by his idealized model. It follows, in theory at least, that Life can simulate any possible computer architecture.

A Turing machine consists of:

- a finite set of symbols;
- a finite set of states;
- an infinite tape on which symbols from the set can be written;
- a read/write head under which the tape can move;

- a program, or set of rules for transforming states, acting on the symbols it reads.

The tape is both input to and output from the machine, and it can read or write only one symbol at a time, and at the same position.

The example below is a very simple application of a set of elementary rules. Further and more detailed discussion can be found elsewhere, for example in the work by Harel [5.11]. In our example, the machine is set to recognize three possible states: A, B, and C and is reading a tape containing symbols 0 and 1 only. Our program works as follows:

```
Repeat
        Read tape
        If state is A
                If symbol is 0
                        move right, set state to B
                If symbol is 1
                        move left, set state to C
        If state is B
                If symbol is 0
                        move left, set state to A
                If symbol is 1
                        move right, set state to B
        If state is C
                If symbol is 0
                        move left, set state to B
                If symbol is 1
                        stop
        Overwrite symbol with 1
for ever
```

It is left to the reader to determine exactly the effect of this program.

5.2.4 Statistics of CAs

Statistics of CAs are concerned with the overall relationships between patterns produced by the same rules at different times.

At one level we are concerned with the local nature of a pattern and how it changes over time. Initially, the pattern may be totally random. We could say that there is no necessary correlation between the states of neighbouring sites. However as a particular rule-set is applied successively over time the element of randomness between neighbours reduces. At the same time the chances of repetition of patterns sequences increases. The precise behaviour depends on the rule set, the set of states, and the number of time steps.

At another level, global statistics are concerned with ensemble properties, in other words the properties of the entire system. Since a cellular automaton is simply a sequence of transformations of a set of values one can borrow techniques from other disciplines. Two particular measures of interest are Hamming distance and entropy, both used extensively in communications and information theory.

Hamming distance is a measure of the number of bits which differ between sequences before and after the application of a transformation. The variation over time is again characteristic of a particular system, but is a global measure which applies to the entire mesh.

Entropy is defined in the field of communications as the information content in data. This concept has been borrowed in turn from statistical thermodynamics, which considers the ensemble behaviour of a gas with a given mean temperature. In this context, entropy indicates the level of disorder in a system. Reversibility of processes is also an important concept in thermodynamics. Irreversible behaviour leads to an alternative definition of the second law, namely that no machine is 100% efficient. Because of the very close similarity between the behaviour of some (irreversible) CAs and a thermodynamic gas, such CA systems are often referred to as lattice gas models. One such case is considered in the next example.

Example 5.5 CA models for the Navier–Stokes equation
The Navier–Stokes equation is an analytical expression of fluid behaviour under a wide range of conditions. If a fluid has a general velocity vector $\mathbf{q}$ then the force acting on the fluid, $m\dfrac{D\mathbf{q}}{Dt}$ equal to density$*$volume$*\dfrac{D\mathbf{q}}{Dt}$ is the sum of the internal and external forces (the operator $\dfrac{D\mathbf{q}}{Dt}$ is the material time derivative). A continuous treatment starts by dividing space into cells, each of dimension Δx, Δy, Δz and quantifying the overall effect in each direction. Thus the force in the x direction is given by

$$\rho\,\Delta x\,\Delta y\,\Delta z\,\frac{D\mathbf{q}_x}{Dt} = F_{\mathbf{x}}\,\rho\,\Delta x\,\Delta y\,\Delta z + P\Delta y\Delta z - (P + \frac{\partial P}{\partial x}\Delta x)\Delta y\,\Delta z.$$

The first term on the right-hand side is the external force. The second term is the internal force (P$*$area$_{yz}$) acting in the x direction. The second and third term together estimate the change in force over the x length of the cell. If this is simplified and if the differences are allowed to tend to zero then we get

$$\frac{Dq_x}{Dt} = \mathbf{F}_x - \frac{1}{\rho}\frac{\partial P}{\partial x}$$

or in general

$$\frac{\mathrm{D}q_x}{\mathrm{D}t} = \mathbf{F} - \frac{1}{\rho}\nabla P \text{ where } \nabla P = \frac{\partial P}{\partial x} + \frac{\partial P}{\partial y} + \frac{\partial P}{\partial z}.$$

Now fluids can be compressible and viscous. The latter property is included by the addition of some extra terms to the equation above. However, since these terms are generally nonlinear, the solution is often very difficult. Some of the problems can be quantified by the use of dimensionless parameters such as the Reynolds number R and the Mach number M. In this case R is a measure of the effects of the nonlinear terms (due to viscosity), while M, which is a ratio of velocities, indicates the effects of compressibility. Although additional effects can be included, this macroscopic treatment can quickly become intractable, particularly for realistic problems such as free convection and weather modelling.

By way of an alternative, the entire problem can be investigated from a microscopic viewpoint where the behaviour of some of the constituent molecules/particles are monitored over time. A cellular automaton approach was first proposed by Hardy, Pazzis, and Pomeau [5.12] and is thus referred to as the *HPP model*. In this case a rectangular mesh is populated by molecules which are subject to a set of rules which describe the motion:

1. Each node can hold up to 4 molecules of equal mass each of which is travelling in any one of the four spatial directions (in 2-dimensions).

2. No two molecules on a node can travel in the same direction.

3. At each time-step a molecule moves to the next node in the direction of its velocity vector.

4. Any configuration of molecules moving in opposite directions and therefore likely to lead to a head-on collision is replaced by another at right angles.

The system is allowed to proceed for some time until the lattice gas approaches equilibrium. At this point, different regions within the model space can be selected and parameters such as mean density and velocity can be determined.

Investigations using this method indicated some problems which were attributed to nonlinear effects which manifest themselves at a macroscopic level in the Navier–Stokes equation. Others [5.13, 5.14] have found that some of these are due to the use of a rectangular mesh. In order to overcome this problem one must either go to three dimensions (with the resultant increase in computational load) or use a triangular mesh. The diagram below (Fig. 5.11) shows some before and after rules for collisions. For head-on collisions two outcomes are equally likely. The diagram also shows rest-state rules which have close similarities with temporary storage in number diffusion, referred to at the end of Section 5.1.

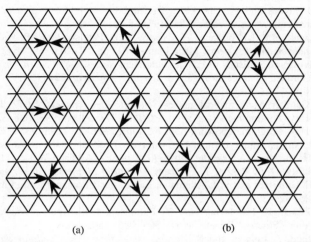

(a)   (b)

Fig. 5.11(a) Collision rules — in each case the initial state is on the left. (b) Rest-state rules.

Other discrepancies between this approach and the conventional Navier–Stokes equation can be removed by adding a small amount of thermal noise to the CA model.

The above rules are applied at every time step. Boundaries are described by having either perfect (specular) reflection or by random scatter back into the model space. It is also possible to have particle generators at one boundary and particle removers at another in the same direction. This simulates a wind-tunnel experiment. After each time step it is possible to estimate the density, ρ and the momentum, ρq within an area using

$$\rho = \sum_i N_i \text{ and } \rho q = \sum_i N_i C_i$$

where C_i is an individual velocity vector.

These basic principles can be extended in many ways. For example, the combustion of fuel in a jet engine would requires a microscopic description of the vaporization and burning of randomly distributed fuel particles, the heat of reaction and energy transfer which increases the velocity of particles. Said et al. [5.15] provide some details of the incorporation of these physical processes into a CA model of dense spray combustion.

References

5.1   J.V. Beck, K.D. Cole, A. Haji-Sheikh, and B. Litkouhi, *Heat conduction using Green's functions*, Hemisphere Publishing Corp. 1992.

5.2   R.M. Bevensee, Probabilistic potential theory applied to electrical engineering problems. Proc. *IEEE* **61** (1973) 423–437.

5.3   J.R. Haynes and W. Shockley, The mobility and life of injected holes and electrons in germanium. *Physical Review* **81** (1951) 835.

5.4   J. Crank, N.R. Mc Farlane, J.C. Newby, G.D. Paterson and J.B. Pedley, *Diffusion processes in environmental systems*, Macmillan (1981) pp. 97–103.

5.5   S. Wolfram, Statistics of cellular automata. *Reviews of Modern Physics* **55** (1983) 601–644.

5.6   M. Gardner, Mathematical games. *Scientific American* **224** (1971) Feb. p. 112, March p. 106, April p. 114.

5.7   M. Gardner, Mathematical games. *Scientific American* **226** (1972) Jan. p. 104.

5.8   W. Hey, Programmers workshop. *PC Plus*, December 1993, 323–325.

5.9   W. Hey, Programmers workshop. *PC Plus*, February 1995, 935–937.

5.10 A.R. Smith, Simple computation—universal cellular spaces *J. ACM* 18 (1971) 339.

5.11 D. Harel, *Algorithmics; the spirit of computing*, Addison-Wesley (1987) pp. 215–254.

5.12 J. Hardy, O. de Pazzis, Y. Pomeau, *Journal of Mathematical Physics* **14** (1973) 1746.

5.13 D. D'Humieres and P. L'Allemand, Lattice gas automata for fluid mechanics. *Physica* **140A** (1986) 326–335.

5.14 U. Frisch, B. Hasslacher, and Y. Pommeau, Lattice gas automata for the Navier–Stokes equation. *Physical Review Letters* **56** (1986) 1505–1508.

5.15 R. Said, S. Loison, and R. Borghi, *Simulation with a cellular automaton of dense spray combustion*, Eurotherm 39, Poitiers, Sept. 1994.

Examples, exercises, and projects

5.1   In Section 5.1.1 we developed an espression for the effect at any position (i, j) in two-dimensional space at time k due to one or more inputs at any other position $(i,' j)$ and time k' This should now be applied to the situation where a temperature input of 100 °C is applied at every time step at $(0, 0)$ and we would like to know the effect at $(2, 2)$ after 4 time steps.

As an extension of this you should consider the case where the temperature at $(0, 0)$ is held at 100 °C for all time. The temperature at $(0, 0)$ due to all other nodes is calculated at time k. The input at $(0, 0)$ for the next time step is then

$$_{k+1}T_{in}(0,0) = 100 - {}_k T(0,0).$$

With a constant temperature input you should now calculate the temperature of position $(2, 2)$ at $k = 4$.

5.2   The Liesegang phenomenon is concerned with the distance and time relationships in periodic precipitations. If a gelatine solution containing potassium dichromate ($K_2Cr_2O_7$) is allowed to set in a glass tube held vertically and if a concentrated solution of silver nitrate ($AgNO_3$) is placed above the jelly then the formation of silver dichromate is observed as bands of precipitate which occur discretely where the following rules seem to apply:

- The distance to the n-th precipitate and its time to formation are related as $\frac{x_n}{\sqrt{t_n}} = C_1$.

- The distances between successive precipitates are related as $\frac{x_{n+1}}{x_n} = C_2$.

If the diffusion constant for $AgNO_3$ is 1.5×10^{-5} and the value for $K_2Cr_2O_7$ is 1×10^{-5} cm^{2}s^{-1} then, for given discretizations Δx and Δt, the reflection coefficient for a biased random walk model is given by

$$\rho = \frac{1}{1 + 2D\frac{\Delta t}{\Delta x^2}}. \quad (D \text{ is the diffusion constant})$$

It should then be possible to develop a one-dimensional model for the in-diffusion of silver nitrate. We can specify a precipitation criterion using the concept of solubility product. If the concentrations of the two species at any node x and time t (designated by square brackets) follows the inequality:

$$[AgNO_3]^2[K_2Cr_2O_7] \geq 2 \times 10^{-7},$$

a precipitate is noted at that node and the local concentrations of the two diffusing species are reset at zero at that node. The simulation then continues. On the basis that the silver nitrate starting concentration is 100 time greater than the dichromate concentration, develop an algorithm which reproduces the Liesegang phenomenon.

5.3 Section 5.1.3 was concerned with shift operators in one dimension and it was observed that a binary sequence was followed, so that at $k = 3$ we have 8 possibilities following an initial unit input from the left: $\rho\rho\rho$, $\rho\rho\tau$, $\rho\tau\rho$, $\rho\tau\tau$, $\tau\rho\rho$, etc. with final positions at $(x-1+1-1)$, $(x-1+1+1)$, $(x-1-1+1)$, etc. Consider how this might be extended to two dimensions. Is it possible to develop a similar set of rules as in the one-dimensional case. Adapt the one-dimensional symbolic code given in the text to give the populations at any position after k iterations.

5.4 Example 5.2 in the text was concerned with the dispersion of smoke from a factory chimney and it was shown that this could by described by addition of a spatially dependent bias to a diffusion model and a scattering equation was presented. How could this equation be modified to include the influence of another drift term which acted at an angle θ to the main flow α?

If a pollutant is added at a constant rate at the centre of a large river whose flow causes a drift bias α, then it should be possible to model this in an identical way to the dispersing smoke. If however the particulate nature of the pollutant causes it to settle under gravity with a vertical bias

β then it should be possible to map the concentration of material on the river bed, assuming that it leaves the simulation once it settles. This general model could be adapted to include: (a) flow velocity which was dependent on the height above the river bed with an α which depended on the square of the river velocity at that height; (b) similar laminar effects in three dimensions to take account of the lack of flow near the river bank, (c) an aggregation effect where the value of β is dependent on the concentration of pollutant at that node.

5.5 Devise a one dimensional CA based on the operation described in eqn (5.8). The space should consist of 10 nodes where node 1 and node 10 are fixed at value 1 at all times. The initial population of the nodes should be a random binary sequence. Observe how the state of order develops when subjected to these boundaries. Investigate the effects of having: (a) one of the boundaries at 1 and the other at 0; (b) both boundaries fixed at zero.

5.6 Consider a one-dimensional CA which starts with a two-dimensional mesh on which the numbers 1 and 0 have been placed with equal probability. The rule is this: take each point (x, y) in turn and take the sum of values at (x, y), $(x+1, y)$, $(x-1, y)$, $(x, y+1)$, and $(x, y-1)$. If the sum is either 4 or less then the value of (x, y) either stays at zero (if it is already zero) or goes to zero (if it is currently 1). Repeat this rule many times and observe the development of an ordered state.

What would happen if the sum had to be 3 (or 5) in order to change the state of (x, y)?

5.7 The diagram of fig. 5.12 shows a disk of ones in a large sea of zeros.

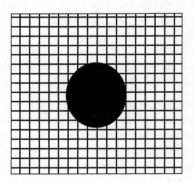

Fig. 5.12 Initial mesh configuration for Problem 5.7.

Start at the bottom left corner and apply the the follwoing rule over the entire mesh during many iterations. If the value of a node, (x, y) is 1, then it will be zero at the next iteration and one of the surrounding nodes (either $(x+1, y)$, $(x-1, y)$, $(x, y+1)$, or $(x, y-1)$) chosen at random will become 1. However, the rule is only applied if the randomly chosen node is currently zero. If this condition is not fulfilled then node (x, y) remains occupied during the current iteration.

5.8 Adapt the game-of-life code in the text to investigate the behaviour of the two initial populations shown in Fig. 5.13.

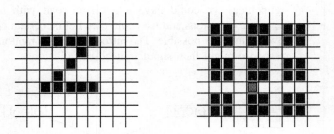

Fig. 5.13 Initial population patterns for Problem 5.8.

In the second case the behaviour of a large population based on the diagram should first be investigated without the shaded entity. This is a 'virus' and its effect on the population is dramatic. What measures could be taken to minimize its effects?

5.9 The flow of traffic along a road can be treated as a one-dimensional lattice gas automaton in the following way. A road can be represented by a one-dimensional mesh of nodes upon which cars are placed at random. Each car can have a different speed within fixed limits and depending on speed a car progresses by a different number of nodes at each iteration. The important factor is the distance (number of nodes) between adjacent cars. We can then define a rate of change of velocity (acceleration/braking) which is given by:

$$\frac{V(n+1) - V(n)}{|x(n+1) - x(n)|} = \Delta V(n).$$

Thus, at the next iteration, the nth vehicle is travelling at $V(n)\pm\Delta V(n)$, depending on whether it was necessary to brake or accelerate. We may define certain limits for $\Delta V(n)$, i.e. it is not possible for any car to have a decelleration $\Delta V(n) = V(n)$ and a more reasonable stopping distance may apply; similarly the acceleration has limitations e.g. 0–60 mph (eventually) for certain makes of cars. Finally, we define a crash as the condition when two cars occupy the same node or if car n would arrive at a position beyond car $n+1$ within an iteration.

Prepare a CA which follows these rules and observe the dynamics for typical random inputs. Adjust the initial conditions so that havoc does not reign. One suggestion may be to have the lead car in a random sequence travel at constant velocity and watch how long it takes the convoy to reach equilibrium. Then arrange for the front vehicle to slow down and watch the propagation of the ripple. If a certain random element were added to $\Delta V(n)$ then we could simulate variable driver response and see what conditions might lead to a pile up. As a further refinement we could have a two-lane road with traffic travelling in opposite directions and we could then define certain circumstances when overtaking was possible. The introduction of a random element in driver response might then simulate head-on collisions.

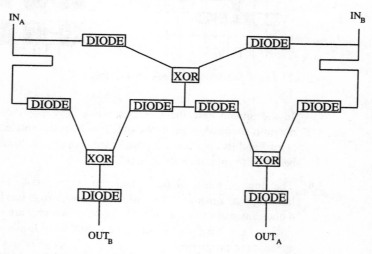

Fig. 5.14 A crossover in *Wires*.

5.10 Consider the design of a crossover in *Wires* which allows one to model the equivalent of a 'flying' connection (IN_A to OUT_A and IN_B to OUT_B) in a real circuit (Fig. 5.14).

The loops near the inputs ensure that signals from IN_A arrive at the same time at the XOR gate before OUT_B and therefore cancel. Thus OUT_B depends only on IN_B (unless signals from the different inputs arrive at the same time).

Probability based models

6

Introduction

There are those who claim that there is no such thing as randomness. They would claim that everything is deterministic even if we don't know what the outcome is. The tossing of a coin is taken to be a random process although it may be possible for a skilled gambler to influence it just as a bowler in cricket can induce spin in the ball. Now in theory it should be possible to model the process. We might know the initial conditions of lift and rotation which are applied to the coin. We might even include the effects of viscous drag of the air. Thus we could calculate the maximum height of the coin as well as the time to land. From this time and a knowledge of the rotational frequency we could calculate the angle at time of landing ($0° \leq \theta < 180° \Rightarrow$ heads, $180° \leq \theta < 360° \Rightarrow$ tails). This gets the coin to the surface and the data could constitute a set of initial conditions with which to model the motion of the coin up to the time it comes to rest. Some example simulations would show that very small variations in initial conditions (e.g. the flapping of a butterfly's wings) could drastically alter the outcome. Thus, after much effort, we might still have a 50/50 chance of getting the right answer. Alternatively, we might leave it to chance. If a process of 'winning' depended on the outcome of a tossed coin, then we would call this a stochastic process. If on the other hand we were confidently able to model the process, then 'winning' would be deterministic and a lot of gambling casinos would be out of business.

We could make observations on a gambler tossing a coin over a period of time and if we noticed that heads appeared more often than tails then we might suspect an irregularity or bias; an individual outcome may appear random, but when examined in the context of previous outcomes, then the belief in random behaviour may be stretched.

This idea of the presence or absence of correlation between outcomes in a sequence of tests is at the core of this chapter where we will be concerned with random sequences. If we could choose a truly random starting point in the number π (3.14159...) then we would obtain a random sequence of digits between 0 and 9. The frequency and amplitude components of genuine 'white' noise have no correlation and this can be utilized in computer generation of random numbers. An analogue-to-digital converter can sample an amplified signal of the noise within some part of a circuit and this can then provide a random binary output of any required byte size.

This chapter is concerned with using random events to develop a range of models, but there is some difficulty because there are several disparate topics which can be included under this heading. In many cases the boundaries between the different subjects are diffuse and therefore the subdivision into headings within this chapter is mainly for convenience.

We will start by considering particles, first one and then many. We will look at the case of a single particle at an instant on a spatial mesh. We will consider the situation where the particle can move at random to land at another mesh point at the start of the next iteration. This will be followed by a discussion of the ensemble (average) behaviour of a large population of particles where each individual can behave in a random way and factors such as population size and number of tests and their effect on accuracy and computational efficiency will be covered. From random particles we then move to random events and several subjects will be covered under the general heading of discrete event modelling.

6.1 Single particle random processes

We start by considering a particle at the centre of a very long discretized bar with equal probability of moving left or right ($P_L = P_R$). Scattering involves the movement of the particle to one of the adjacent sites. In a two-dimensional mesh the probabilities of moving up, down, left or right are equal ($P_U = P_D = P_L = P_R = \frac{1}{4}$).

We can look at the behaviour of a single particle many times or we can look at an ensemble of particles at the same time. Thus, in a one-dimensional problem involving say 1 000 000 particles, we can assume that 500 000 move to the left and 500 000 move to the right. For reasons of computational efficiency we ignore the fact that each time the experiment is repeated the number to left and right will be different (although approximately equal). The situation where the behaviour of each random particle is charted and an average then calculated is the subject of Section 6.2.

We could also have a range of circumstances where the probabilities summed to unity but were not equal, i.e. one direction has a bias. The case of equal probabilities is referred to as *simple random walk* (SRW) and the other is called biased or *correlated random walk* (CRW). Because it is possible to use SRW and CRW models to treat a wide variety of problems, these two topics are discussed separately.

6.1.1 Single particle random motion with symmetric scatter

If a one-dimensional static thermal problem has 0 °C as one boundary and 100 °C as the other, then we estimate the temperature at any intermediate point P using the *drunkard's walk*, a single particle/many repetition technique. The space between the boundaries is discretized into a series of nodes. The *walker* starting from any node has only two options, either to turn left of to turn right. If the walker does not encounter a boundary then it is allowed to repeat the left/right choice. If a boundary is encountered, then the value of the boundary (0 °C or 100 °C) is added to an accumulator and the walker is returned to the origin P for another test. The value of temperature at P is then the mean of many tests.

The Turbo C code given below is for an equivalent electrostatics problem where the point P is located at 40% of the total length with respect the ground (0 V) electrode. The other electrode is at 100 V.

```c
#include "stdio.h"
#include "time.h"
#include "stdlib.h"
#define SIZE 10                    /*  Size of array */
#define START 4                    /*  starting  position  */
#define ATTEMPTS 30000             /*  less  than  32767  !  */

void  main(void)
{
    char now;                      /*  present  position  */
    int total_score=0;             /*  number  of  'wins'  */
    int hit_counter=0;             /*  number  of  boundaries  hit  */
    int power=1;                   /*  for  logarithmic  scale  printing  */
    now=START;
    /* set the random number seed */
    randomize();
    /* start of output */
    printf("---  random  walk  ---");
    /* loop till ready */
    while(hit_counter <= ATTEMPTS)
    {
        /* calculate next position */
        if(rand()  %  2)
            now++;
        else
            now--;
        switch  (now)
        {
        /* if 'win' add 1 to total score */
        case SIZE:
            total_score  +=1;
        /* if boundary hit reset present position to start */
        /* increment  hit  counter  */
        /* and if print flag set, output latest score */
```

```
    case 0:
      now=START;
      hit_counter++;
      if(hit_counter==10 || hit_counter==100 ||
            hit_counter==1000 || hit_counter==10000)
        power*=10;
      if(!(hit_counter%power))
      {
          printf("\n%5d%8.1f",hit_counter,(100.0  *  total_score)/hit_counter)
      }
   /* otherwise continue walking */
   default:;
   }
 /* end loop */
 }
}

}
```

The code allows the test to be repeated 30 000 times. This is because the accuracy of the estimate depends on the square root of the number of tries. The value of the potential at P as a function of the number of tries up to 10 000 for one particular run of the code is shown in Fig. 6.1 and indicates how the estimate of the potential of P converges towards 40 V. It should be noted that if the code were to be rerun, then the initial scatter of values would be different although the ultimate result would be the same.

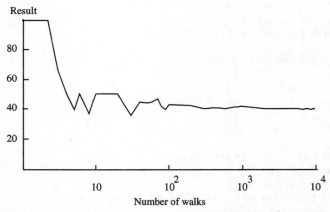

Fig. 6.1 The estimated value as a function of the number of tries in a one-dimensioned simple random walk model.

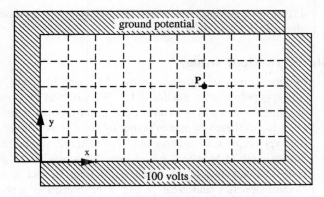

Fig. 6.2 A series of equidistant mesh points in a charge-free region surrounded by two electrodes.

The same technique can be applied in two or three dimensions to obtain a solution for the Laplace equation. The particular example comprises a closed region of free space bounded by electrodes as shown in Fig. 6.2.

Note that this is almost identical to Fig. 3.2 except that we are considering a single point P. A particle is located at P and, using navigational notation, is allowed to move either north, south, east, or west at every iteration. After several steps it will arrive at a boundary and the value of the potential of that boundary will be added to the store. The particle is then returned to the starting point and the process is repeated for a total of N_T experiments. The potential of the point P is then estimated as:

$$V_P = \frac{\sum V_{B_i}}{N_T}. \tag{6.1}$$

This is obviously computationally inefficient as the same type of experiment would have to be repeated for every point in the mesh. However there is an adaptation of this which considers many particles and is much more efficient. If we assume that we have a very large number of particles at a point in a two-dimensional mesh at the start of an experiment and if these particles are allowed to move under simple random walk rules, then we could say that for sufficiently big populations one quarter would move to each of the four adjacent nodes. At the next time step this process would be repeated and as the number of iterations increased, so the number of populated points would also increase. Indeed the number of particles occupying sites would be a Gaussian distribution with the point P as centre. It will be noted that this description is identical to the finite-difference method which was a consequence of eqn (3.28). For one-dimensional problems, simple random walk (number diffusion), finite difference and TLM ($\rho = \tau = 1/2$) models for diffusion are identical in all respects.

6.1.2 Biased random walk

In this section we present ideas which are in many ways similar to the concepts presented in Section 5.1.3. We will start by considering an assemble of particles which is not constrained to have equal transition probabilities; in other words there is a bias in favour of one or other direction. We can consider two types of bias: local and global. Local bias is similar to the situation in one-dimensional TLM where $\rho \neq \tau$. It is known in TLM that this may lead to faster convergence of an iterative solution. In fact it is possible to ignore electromagnetics and treat ρ and τ as transition probabilities. Using this concept we can see a local transition as one which relates to the particle assembly itself. Thus, if a quantity of particles is incident from the left, it will have transition probabilities ρ and τ so that these represents the relative fraction which are reflected and transmitted. If the controlling factor is the interaction between particles and the local environment then an assembly incident from the right might also have the same transition probabilities as is shown in Fig. 6.3.

These ideas can be developed using concepts from Chapters 4 and 5. If we were to use the transition probabilities to estimate the concentration of a diffusing species, then we could get expressions in terms of left- and right-moving probabilities. Supposing the overall concentration of species at position x at time k was $_kC(x)$ then the left- and right-moving components could be eliminated to yield:

$$_{k+1}C(x) = \tau[_kC(x+1) + _kC(x-1)] + (\rho^2 - \tau^2)_{k-1}C(x). \tag{6.2}$$

There is much in common with finite difference expressions for similar problems (see Chapter 3); indeed, when $\rho = \tau = 0.5$ the expression is identical with eqn (3.45). However, in all other circumstances, the third term indicates that there is a correlation between successive time steps and for that reason the process is often called *the correlated random walk* (CRW).

Fig. 6.3 Locally biased transition probabilities.

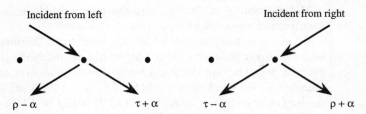

Fig. 6.4 Globally biased transition probabilities. The values of ρ and τ may be equal or there may also be a local bias.

Global bias is identical to the situation described by Example 5.2. In this case there is an external influence which causes a bias (drift) with respect to the mesh itself. This is shown in Fig. 6.4.

These two subsections have served to highlight the probabilistic descriptions of topics which have been treated in detail in earlier chapters. The next subsection represents a novel interpretation of probabilities.

6.1.3 Local interaction simulation approach (LISA)

This method, which has been developed by Delsanto and colleagues at the Polytechnic of Turin, starts with a walker which can only move to an adjacent mesh point at each time step. From a knowledge of the transition probability between mesh points one can estimate the occupational probability for any mesh point at any subsequent time.

In the first instance this is best understood using a one-dimensional mesh. This is shown for subsequent time steps in Fig. 6.5. We define $P(x)$ as the probability of finding the walker at point x and α is the probability of transition to an adjacent node. Thus $(1 - 2\alpha) P(x)$ is the probability of finding a walker who started at x still there at the next time step.

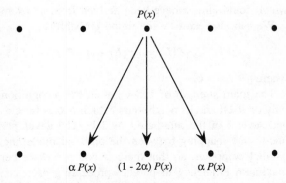

Fig. 6.5 One-dimensional transition probabilities in a simple LISA formulation.

The probability of finding something at x is $(1 - 2\alpha)\ P(x)$ plus the contributions from neighbours, $\alpha P(x - 1)$ and $\alpha P(x + 1)$.

We can also define regions which Kaniadakis and Delsanto [6.1] call *limbo* where a walker leaves the mesh with transition probability β. So, if we had a mesh of N points with absorbing boundaries at either end, then the transition probabilities for $x = 1$ or N would be $(1 - \alpha - \beta)P(x)$ and the probability of the walker moving into limbo from 1 or N would be $\beta P(x)$.

The rates of change of probability at x during a unit time step at iteration k is

$$(1 - 2\alpha)P(x) + \alpha P(x - 1) + \alpha P(x + 1) - P(x).$$

This and the other rates can be summarised as:

$$\frac{d_k P(x)}{dt} = C[_k P(x - 1) + _k P(x + 1) - 2_k P(x)] \quad (1 < x < N) \tag{6.3a}$$

$$\frac{d_k P(1)}{dt} = C[_k P(2) - _k P(1)] - \gamma_k P(1) \tag{6.3b}$$

$$\frac{d_k P(N)}{dt} = C[_k P(N - 1) - _k P(N)] - \gamma_k P(N) \tag{6.3c}$$

where $C = \alpha/\Delta t$ and $\gamma = \beta \Delta t$ are the measurable transition rates to neighbouring sites and to limbo respectively.

The rate of change of $P(x)$ can be approximated by:

$$\frac{_{k+1}P(x) - _k P(x)}{\Delta t}. \tag{6.4}$$

Using eqns (6.3) we then have the following relations:

$$_{k+1}P(x) = \alpha[_k P(x - 1) + _k P(x + 1)] + (1 - 2\alpha)_k P(x) \ (1 < x < N)$$
$$_{k+1}P(1) = \alpha_k P(2) + (1 - \alpha - \beta)_k P(1)] \tag{6.5}$$
$$_{k+1}P(N) = \alpha_k P(N - 1) + (1 - \alpha - \beta)_k P(N)]$$

which could almost have been derived from an examination of Fig. 6.5.

We can also estimate the limbo probability

$$_{k+1}P(L) = _k P(L) + \beta[_k P(1) + _k P(N)] \tag{6.6}$$

where $_0 P(L) = 0$.

The main strength of LISA lies in the recognition that each cell depends only on itself and its neighbours, which makes for close similarities with TLM and several of the rule-based models. The developers of LISA have used a one-to-one mapping between the physical modelling space and a massively parallel computer, so that there is one processor per spatial node. Now for maximum efficiency in parallel processing each computer should have the same task to perform at each time step. Accordingly equations (6.5) and (6.6) are recast as

$$_{k+1}P(x) = \alpha[_kP(x-1) + {}_kP(x+1)] + \eta(x)\,{}_kP(x), 1 \leq x \leq N, \qquad (6.7)$$

where $\eta(x)$ is a coefficient which is input as initial data and may vary arbitrarily.

The spatial variation of probabilities can also be expressed in finite difference form so that the second derivative can be given as:

$$\frac{\partial^2{}_kP(x)}{\partial x^2} = \frac{{}_kP(x+1) + {}_kP(x-1) - 2{}_kP(x)}{\Delta x^2}. \qquad (6.8)$$

Thus we can get a probabilistic expression for the diffusion equation

$$\alpha[_kP(x-1) + {}_kP(x+1)] + \eta(x)\,{}_kP(x) = D\frac{{}_kP(x+1) + {}_kP(x-1) - 2{}_kP(x)}{\Delta x^2} \qquad (6.9)$$

which leads to the well-known expressions for the diffusion constant

$$D = C\,\Delta x^2 \quad \text{or} \quad D = \alpha\,\Delta x^2/\Delta t.$$

This can be used to define the fundamental timestep $\Delta t = \alpha \Delta x^2/D$.

Example 6.1 One dimensional diffusion with end losses
The above equations can be applied to a one-dimensional rod consisting of a large number of nodes where the initial occupational probability of all except one (the source) is zero and the occupational probability of the

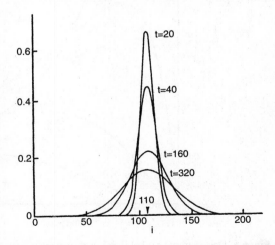

Fig. 6.6 Occupational probabilities at various times following a single initial unity population at the centre of the one-dimensional problem. The vertical axis is occupational probability of any node at any iteration. The horizontal axis is node number.

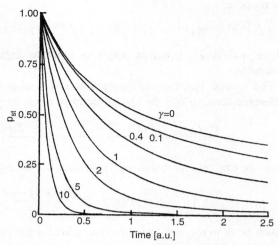

Fig. 6.7 The influence of γ on the occupational probability at the source node as a function of time at various times (in time-step units) following a single initial unity population at the centre of the one-dimensional problem.

source node is unity. For simplicity the constant C is taken as unity. With the source placed at the centre of the rod one gets a Gaussian diffusion as shown in Fig. 6.6. The effect of different limbo transition probabilities can easily be estimated and this is shown in Fig. 6.7. Figure 6.8 shows the

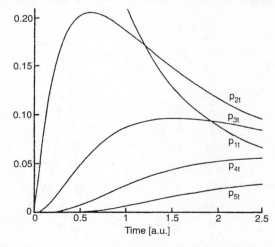

Fig. 6.8 Occupational probabilities at different nodes as a function of time (with γ = 1) following a single initial unity population at the centre of the one-dimensional problem. The annotations next to the curves refer to the node locations with respect to the source.

occupational probabilities at different sites with $\gamma = 1$. These figures are taken from Kaniadakis and Delsanto [6.1] who also describe how the above principles can be applied to two dimensional problems [6.2, 6.3].

Figures 6.6, 6.7, and 6.8 reprinted from *Mathl. Comput. Modelling*, 17(10), G. Kandiakis, P.P. Delsanto, and C.A. Condat ' Local interaction simulation approach to the solution of diffusion problems, 1993, pp 31–42. Reprinted with kind permission of Elsevier Science Ltd.

6.2 Ensemble properties of multi-particle random processes

This sections considers particles and processes where every event is randomly determined. It starts with a discussion of simple Monte Carlo processes where information can be derived by 'averaging' over many tests. We then go on to outline where these ideas can be used in practical models of physical processes.

6.2.1 Simple Monte Carlo processes

Example 6.2 Buffon's method (1777) for estimating π

Let us start with a large surface comprising of parallel lines which are separated by a distance $2a$ (e.g. a very large planked floor). If a needle of length $2b$ (where $b \le a$) is dropped onto this floor, then we could observe whether the needle lies across a line (joint between planks) or entirely within a single plank. If during a number of tests (N_T) it was observed that there were n occasions when the needle lay across a line then it would be noted that as N_T got larger, the ratio n/NT would become constant. This is in fact a measure of the probability p that the needle will cross a line. It can be shown that

$$p = \frac{2b}{a\pi}.$$

Thus, as $N_T \to \infty$, the value of π can be determined.

Random number generation can also be used in Monte Carlo integration. Suppose that it is required to integrate a function $f(x)$ in the range $x = 0$ to $x = M$, then we could define a two dimensional ($N \times M$) mesh which bounded the function as shown in Fig. 6.9. It is of course assumed that the mesh is fine enough so that the discretization of the curve does not introduce appreciable errors.

A random number generator produces pairs of numbers (i, j) in the range $0 \le i \le M$ and $0 \le j \le N$. These are then treated as follows.

If $f(i) - j$ is zero or positive then the pair (i, j) lies on or below the curve and this *success* is recorded, otherwise the *failure* is ignored. Another pair is

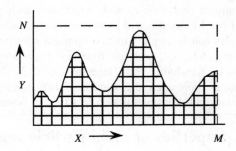

Fig. 6.9 A function f(x) within an ($N \times M$) rectangle.

then produced and the exercise is repeated. Assuming that the area of the rectangle is $N \times M$, then if the total number of random (i, j) pairs was 50 000 and if 30 000 of these lay on or below the curve, then we could say that the area under the curve was:

$$\frac{30\ 000}{50\ 000}(N \times M).$$

If the experiment were repeated the results would not be precisely the same although it would be close to the above value and of course the larger the number of random pairs that were tested, the more accurate would be the estimate of the area under the curve. Nevertheless, it must be remembered that this estimate is not the integral of the function $f(x)$ in the range $0 \leq x \leq M$, but the integral of the discretized function which represents $f(x)$.

Example 6.3 Monte Carlo solution for the Laplace equation
Let us suppose that we have a closed system consisting of four electrodes, each with a different voltage, then it is possible to determine the potential, $V(x,y)$ of the enclosed space point by point using random number generation. The point (x,y) has four construction lines drawn to the intersections of the electrodes as shown in Fig. 6.10

It can be seen that the chosen point (x, y) lies close to V_C but a long way from V_A, while it lies equidistant from V_B and V_D. We might therefore intuitively expect it to have a large component of V_C, lesser, but near equal components from V_B and V_D and a much smaller contribution from V_A.

The space enclosed by the electrodes is discretized into a mesh. Random pairs of numbers (i, j) are produced which lie within this area. A check is then made to identify the triangle within which the pair fall. Thus, if the random (i, j) was bounded by the triangle which had V_A as a side and (x, y) as the opposite apex, then a *win* for V_A would be recorded. This process is

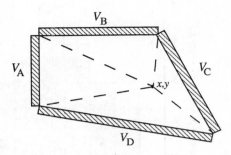

Fig. 6.10 Electrode system to be modelled using a Monte Carlo method.

repeated for a large number, N_T of pairs and has *wins* N_A, N_B, N_C and N_D (where $N_A + N_B + N_C + N_D = N_T$). The calculation of potential can best be outlined by starting with a one dimensional problem (e.g. the finite space between a pair of infinitely long parallel rods). The potential $V(x)$, where x lies between electrode V_A and V_B could be calculated from the *wins*, N_A and N_B as:

$$V(x) = \frac{N_B V_A + N_A V_B}{N_A + N_B}.$$

Obviously, if x lay much closer to V_A then to V_B, then N_A would be very much less than N_B and so $V(x) \approx V_A$.

If a space is bounded by three electrodes so that the Monte Carlo test produces wins N_A, N_B, and N_C, then the potential is given by:

$$V(x,y) = \frac{N_B N_C V_A + N_A N_C V_B + N_A N_B V_C}{N_A N_B + N_B N_C + N_C N_A}.$$

An expression for the potential of the point in the quadrilateral problem shown in Fig. 6.10 can be obtained by extension and this process can be repeated for every point in the enclosed space.

6.2.2 The extension of Monte Carlo models

We can very quickly become immersed in heavy probability and statistics, but it is not the purpose of this book to do so; rather we aim to provide a progressive introduction to a range of concepts.

It must be remembered that in a Monte Carlo model of a real system such variables as particle velocity will be stochastic. In simple terms, if we were to consider a charge carrier inside a piece of semiconducting material at the instant after a collision event, then we would not be able to accurately predict

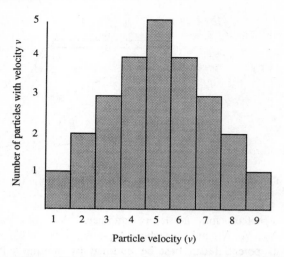

Fig. 6.11 A simple particle/speed distribution for a hypothetical one-dimensional Monte Carlo simulation of particle dynamics.

its velocity. However we could make some statistical observations, and if these were used in a suitable model, then we might be able to draw sensible conclusions.

We will now consider a simple one-dimensional example where a collection of particles are initially located at position x. For the purpose of this example we will assume that there are 1000 particles and their velocity can be described by the histogram shown in Fig. 6.11. We could define a fundamental time discretization such that particle with velocity '1' would travel a distance Δx and a particle with double the velocity would travel twice the distance and so on. We could also assume that any particle would have an equal chance of moving in the $+x$ or $-x$ direction.

The simulation of this problem could use a coin and a 25 point roulette wheel. This point would have one point numbered 1 and one point numbered 9 (velocities 1 and 9). There would be two points each with 2 and 8, three points with 3 and 7 and so on. One particle in the ensemble is selected and a coin is tossed to determine motion left (heads) or right (tails). The wheel is rolled. If the result was velocity $= 4$, direction $=$ left then the particle would be moved to position $(x - 4)$. This process (which could be easily done by computer) is repeated for every particle. It could then be repeated over many time steps for every particle and the outcome at any time could be plotted as a particle distribution.

So far, the particles are completely free moving and not subject to constraints. The situation could be altered by placing a perfectly reflecting boundary at a position, e.g. $(x + 10)$. This boundary would conserve momentum, so that if it was hit by a particle starting from $(x + 8)$ and moving with velocity 5, then the resting place for the particle at the end of the time step would be $(x + 7)$.

These ideas can be easily extended to two dimensions and, instead of the velocity distribution shown in Fig. 6.11, the particles could have a Maxwell – Boltzman distribution, which is a characteristic of real gas particles. Thus we could start to think of the filling of a void by gas-like particles. In the kinetic theory of gases it is assumed that:

- particles are in continuous random motion;

- particles occupy negligible volume;

- particles do not interact with each other.

Thus we could have an ensemble of particles injected into a container, as shown in Fig. 6.12 with a velocity (to the right) $v_d + v$, where v is a Maxwellian distribution of velocities (this even allows for some particles which might be moving against the flow).

At first there is nothing within the container to cause collisions, so that a particle which lands at $(x, 0)$ will continue at the same velocity and in the same direction at the next time step. If however, a 'slow' particle is occupying a position and a faster particle attempts to occupy the same location, then a collision will occur. Normally, the total momentum will be conserved, and the scattering will be just like two billiard balls. Over a range of observations the angle θ of the incident ball might be expected to have any value from $0°$ to $360°$ while the 'shunted' particle would be expected to move in the x direction. In order to allow for variability, the momentum exchange between particles and the resultant scattering angles would be chosen at random.

These ideas are realized in a simple way in the code which is shown below. Twenty particles, each with a velocity of 20 arbitrary units in the x direction are introduced into a region. The subsequent scattering at discrete intervals is presumed to introduce a simple exponential change to the present velocity of each particle (50% chance of no change, 25% chance of a ± 2 arbitrary units

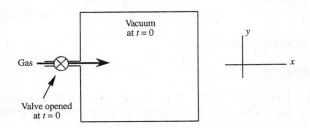

Fig. 6.12 Gas injection into a void showing the computational axis.

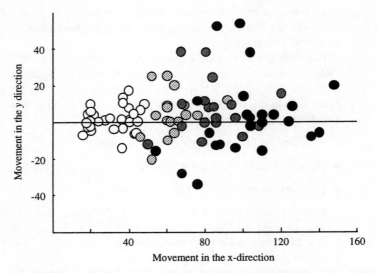

Fig. 6.13 The motion during 5 time steps of 20 particles injected into an empty space with an initial velocity $v_x = 20$, $v_y = 0$. The particle conditions at succeeding time steps are designated by five progressively darker levels of shading.

change, etc.) and the results over 5 time steps, shown in Fig. 6.13, confirm the way in which the particles will disperse.

```
#include "stdio.h"
#include "stdlib.h"

#define SIZE 20              /*  Size of particle array */
#define Vzero 20             /* starting x velocity */
#define STEPS 6              /* number of time steps (including 0) */

int  get_increment(void);

void  main(void)
{
    int  dist_x[SIZE][STEPS];   /* array of x-distances travelled at each step */
    int  dist_y[SIZE][STEPS];   /* array of y-distances travelled at each step */
```

```c
    int size;                    /* current  particle  within  array */
    int step;                    /* current  time  step */

    randomize();
    printf("\nPositions of each particle at each time step\n");
    /* initialise distances travelled for time step 0 */
    for  (size=0;size<SIZE;size++)
    {
        dist_x[size][0]  =  Vzero;
        dist_y[size][0]  =  0;
    }
    /* calculate distances travelled for each time step */
    {for  (step=1;step<STEPS;step++)
        for  (size=0;size<SIZE;size++)
        {
            dist_x[size][step]  =  dist_x[size][step-1]  +  get_increment();
            dist_y[size][step]  =  dist_y[size][step-1]  +  get_increment();
        }
    }
    /* calculate and print total distance travelled at each time step */
    for(size=0;size<SIZE;size++)
    {
        printf("\n%2d        (%3d,%3d)",size+1,dist_x[size][0],dist_y[size][0]);
        for(step=1;step<STEPS;step++)
        {
            dist_x[size][step]+=dist_x[size][step-1];
            dist_y[size][step]+=dist_y[size][step-1];
            printf("        (%3d,%3d)",dist_x[size][step],dist_y[size][step]);
        }
    }
}

int  get_increment(void)
{
    int  sign;
    int  increment;

    if(rand() % 2) sign = 1; else sign = -1;
    increment = rand() % 64;
    if (increment < 32) return 0;
    if (increment < 48) return sign * 2;
    if (increment < 56) return sign * 4;
    if (increment < 60) return sign * 6;
    if (increment < 62) return sign * 8;
    return sign * 10;
}
```

Instead of the situation shown in Figs. 6.12 and 6.13, we could have a nozzle injecting particles (at a fixed rate and velocity distribution) into a region which already contained a random distribution of particles. In this manner we could simulate the injection and dispersal of smoke. Alternatively, we could treat the incoming particles of finite volume as reactants (such as oil particles) entering a combustion chamber, have a chemical reaction taking place at the surface of each particle so that the over a period of time the volume of remaining (uncombusted oil) decreases. This could be the start of an investigation into the effects of particle size or dispersion on the speed of combustion.

6.3 Discrete event models

This section uses probability to model systems where some significant event may occur at random. The classic case is queuing.

Customers are passing through a supermarket at the rate of 180 per hour. There are currently 6 check-outs which take on average 2 minutes to process a customer's purchases. When business is slack customers will arrive at a check-out at random. However, when there is a queue, customers may choose the shortest queue or the queue where those in front apparently have the smallest number of purchases. The faster that customers can be processed, the larger the income (and customer satisfaction will ensure future business). However, providing an additional check-out has cost implications. A supermarket manager will therefore wish to know at what point this becomes an economic proposition.

Although much can be said about average behaviour of customers, it is very useful to model this for a large population over a large number of random events. This will be covered in greater detail in Section 6.3.2 and will be extended to traffic models in Section 6.3.3. Firstly we will consider a discrete event problem from the semiconductor industry.

6.3.1 Modelling of electronic component yield

The manufacture of semiconductor devices generally require many process steps [6.4]. At an early stage it will be necessary to diffuse junctions. In integrated-circuit fabrication these may be located at small depths below the surface. For power devices (e.g. thyristors) the depths may be orders of magnitude greater. It may be surprising to realize that in spite of the strictest process control the devices which are processed on the same silicon wafer will have slightly different junction depths. If as a rough guide we assume that 80% are within specification, then we can say that there has been an 80% yield. Now, supposing that there were three further process steps before a complete device was produced and if each step had an 80% yield, then after the second step we would have 80% of the original 80%, i.e. 64%, would now be within specification. This would mean that the overall yield would be 40.96% which is not at all good. Except to indicate the importance of maintaining a high yield at every step, we do not get much information about the performance of the final devices. An alternative approach is required and the following, which is due to Kennedy [6.5] uses a discrete random event approach to do the work which is normally undertaken by a team of R & D engineers.

Let us imagine that a semiconductor device requires the following critical process steps:

- diffusion of a junction (p-type);

- growth of a masking oxide layer;

- photolithography and etching of pattern in oxide layer;

- diffusion of a junction (n-type);

- growth of a masking oxide layer;

- photolithography and etching of holes for electrical contacts;

- metallization and formation of electrical interconnects.

Each of these steps has a spread, i.e. a junction (nominally 0.5 μm below the surface) will be anywhere in the range 0.4–0.6 μm. Because of the nature of the process (and as a result of extensive measurements) it is known that the distribution is normal with a mean of 0.5 μm. This is shown in Fig. 6.14.

There is a similar set of distribution curves for every other process step and any finished device will be the result of a random sequence of discrete events within these distributions.

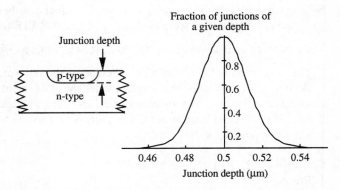

Fig. 6.14 A pn junction with normal distribution of the diffusion depth.

The simulation of the fabrication process can then be automated using a random number generator which chooses a junction depth for step 1 according to the relevant distribution. This information is then carried over to the next step and a random oxidation thickness is obtained. Once the remaining steps have been randomly chosen within the distribution envelopes the product will be a device with a set of process features which can then be used to calculate the electronic performance. Provided the fabrication simulation is repeated for a sufficiently large number of devices it is possible to derive a series of distributions of the required performance parameters, just as if a development engineer had processed many batches and obtained experimental results. These may indicate that the deleterious influences of process spread (previously termed as yield) may counteract each other (highly desirable) or may compound to result in a very low yield.

6.3.2 Queuing and GPSS-type models

GPSS (General Purpose Simulation System) is a software system for the dynamic simulation of stochastic problems. It is in fact a formalized language within which the Kennedy treatment of semiconductor yield could be described and treated. It is not our intention to present this section as a GPSS primer. Instead the plan is to use a GPSS-like syntax to show how discrete event models can be developed into a programming language which can be used as a general modelling tool. As an example we could consider the large supermarket mentioned previously. It currently has N check-outs and at any time it will close one down temporarily and redeploy staff stocking shelves if there are less than two people queuing at each. The customer tolerance has been well characterized and management wish to know the effects of operating only $N-1$ check-outs under normal circumstances. To get some way towards tackling such problems we will need to define action words which together constitute a modelling syntax.

The first of these is 'GENERATE (statistics)'. This means that an event will be generated at random according to some statistical distribution over time. If a rectangular distribution were assumed then GENERATE (a, b) might mean:

the average arrival time is a, but the spread is from $a - b$ to $a + b$.

The next word is 'TERMINATE'; the end of the process.

With these two words one might model the movement of cars past an observation point on a motorway, e.g. cars pass at an average rate of one very 3 secs. However the bounds are 1 per second to 1 every 5 seconds and the process is repeated for 100 cars:

GENERATE (3,2) *(i.e. 3+2 to 3−2)*
TERMINATE
START 100

If it was intended to simulate traffic movement over a period of time which included rush-hour, then the statistics of GENERATE could be altered accordingly over time.

Taking this example further we could consider the arrival of cars at an observation point and their movement through a section of motorway. The word 'ADVANCE (statistics)' uses the reside distribution to create a range of times before the cars exit from the section

```
GENERATE (statistics)
ADVANCE (statistics)
TERMINATE
START 100
```

We can now move one step further by having a toll house at some point within the section. Only one vehicle can pass through this at any time and there is a distribution of service times. This can be characterized by a pair of words 'SEIZE (object)' and 'RELEASE (object)', which can create a queue since no event is able to move from the GENERATE block until the previous event has been processed by the server (toll in this case):

```
GENERATE (statistics)
ADVANCE (statistics(road section A))
SEIZE (toll)
ADVANCE (statistics(toll))
RELEASE (toll)
ADVANCE (statistics(road section B))
TERMINATE
START 200
```

The statistical analysis of this could give information about the average reside times in each section, the average size of queue, and the utilization of the toll. This example is in fact identical to a customer who enters a shop, browses, pays for purchases at a single check-out, and then browses before leaving the shop.

We also have to consider series and parallel arrangements. Traffic travelling along a toll road will arrive at many toll gates (servers in parallel), while a large supermarket will have parallel queues with parallel servers.

The serial case can have important industrial model applications, particularly if a manufacturing process involves several sequential steps (not unlike the case in Section 6.3.1) and the statistics could be used to identify bottle-necks which might be economically eased by the use of parallel operations.

For parallel servers with a single queue we could think of the immigration controls in many international airports. If there were three immigration officers working in parallel then we would need to define a store with an enter and leave block

```
STORE(3)
GENERATE (statistics)
ENTER (store)
ADVANCE (statistics)
LEAVE (store)
TERMINATE
```

In many real examples the servers in parallel might be subject to the use of a conditional statement. *If* one server is occupied go to another server currently not occupied. Thus the queuing at a road toll with 2 gates might be simulated as:

```
GENERATE (statistics)
IF (toll 1 occupied, toll2)
ARRIVE (toll 1)
SEIZE (toll 1)
ADVANCE (statistics(toll))
RELEASE (toll 1)
LEAVE (toll 1)
TERMINATE
ARRIVE (toll 2)
SEIZE (toll 2)
ADVANCE (statistics(toll))
RELEASE (toll 2)
LEAVE (toll 2)
TERMINATE
```

The analysis would include details about the average queue size and the utilization of the toll gates. A small modification could deal with toll operators who had slightly different client handling times by having different toll statistics in each case.

Such systems also have provision for *go to* statements, which can be unconditional or statistical. The unconditional case can be described by considering a factory with several workers who have access to only one piece of equipment. The classic example is several potters who can throw the clay within a time interval. Each piece of clay must then be fired in the oven. There is only one oven which is available to only one user at a time and the firing times vary within a known distribution. Once a piece is fired a worker then returns to throw some more clay.

```
GENERATE (statistics,N) *(N-workers in parallel)*
begin ADVANCE (statistics(clay))
SEIZE (furnace)
ADVANCE (statistics(firing))
RELEASE (furnace)
GOTO (begin)
```

This can be used to determine the optimum value of N, i.e. that which ensures the highest utilization of each worker.

The statistical 'go to' has two possible outcomes where a probability is defined as a parameter. Thus, GOTO(location,0.1) would result in a one in ten chance of being sent to 'location'. The process of visiting an out-patient clinic in a hospital might be as follows:

```
GENERATE (statistics)
ARRIVE (receptionist)
SEIZE (receptionist)
ADVANCE (statistics(receptionist))
RELEASE (receptionist)
LEAVE (receptionist)
GOTO (nurse,0.75)
TERMINATE
ARRIVE (Nurse)
SEIZE (Nurse)
ADVANCE (statistics((Nurse))
RELEASE (Nurse)
LEAVE (Nurse)
TERMINATE
```

Minor modifications could be added to include possible attention by a doctor as well as possible returns to the receptionist for the arrangement of future appointments etc.

For a more detailed introduction to this subject, the reader is recommended to consult the text by Stål [6.6].

References

6.1 G. Kaniadakis, P.P. Delsanto and C.A. Condat, *A local interaction simulation approach to the solution of diffusion problems. Mathematical Computer Modelling* **17** (1993) 31–42.

6.2 P.P. Delsanto, G. Kaniadakis, M. Scalerandi and D. Iordache, Time scaling in the processing simulation of diffusion processes. *Computational Mathematical Applications* **27** (1994) 51–61.

6.3 P.P. Delsanto, G. Kaniadakis, M. Scalerandi and D. Iordache, A renormalisation group approach to the simulation of diffusion problems. *Maths. Comp. Mod.* **19** (1994) 1–8.

6.4   D. de Cogan, *The design and technology of integrated circuits*, John Wiley and Sons, Chichester 1990.

6.5   D.P. Kennedy *Monte Carlo analysis of bipolar transistor fabrication process*. Proc. NASECODE-II, Dublin 1981, Boole Press, pp. 63–87.

6.6   I. Ståhl, *Introduction to simulation with GPSS*, Prentice-Hall (UK) 1990.

Examples, exercises, and projects

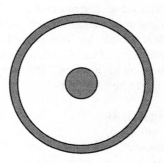

Fig. 6.15 Section of cylindrical oven.

6.1   The diagram in Fig. 6.15 shows two concentric cylinders whose radii are in the ratio 1:4. The inner one is held at 1000 °C and the outer at 900 °C. Use a drunkard's walk approach to estimate the temperature gradient at the midpoint between the two cylinders.

It will be worth comparing the results using Cartesian and polar discretizations. In the case of the Cartesian discretization it will be worth seeing whether there is any difference in the estimated gradient along the x axis and along the diagonal ($\theta = 45°$) axis.

What steps would need to be taken in a model of this type if the two cylinders were eccentric?

6.2   Use a Monte Carlo approach to estimate the area within a semi-cardioid which is given by $r = 0.5 \left[\cos \theta + 1 \right]$ in the range $0 < \theta < 2\pi$

6.3   In Section 4.2.4 we considered models for underwater acoustic propagation and the results for the Buckingham and Tolstoy wedge were shown in Fig. 4.15. In this problem we use TLM to simulate the effects of sea surface waves on sound propagation. The structure to be modelled is shown in Fig. 6.16. The source, which generates a 25 Hz signal is located 50 nodes below the sea surface. There are two receivers. R_1 is 40 nodes below the surface and R_2 is 60 nodes below the surface. The left- and right-hand boundaries, which are a long distance from the centre of action have $\rho = 0$ for convenience.

Once this part of the model has been constructed we can then move to have a wave-like sea surface boundary. As a first estimate we can

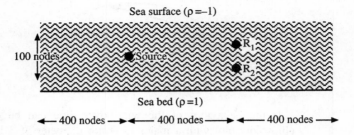

Fig. 6.16 Shallow-water sound-propagation scenario.

arrange to have a discretized sinusoidal boundary which has a wavelength of 10 nodes and a peak-to-peak amplitude of 10 nodes. It might also be worthwhile to use the ideas presented in Problem 4.3 to define a continuous rather than a discretized boundary.

Within the time frame of an acoustic simulation the surface boundary can be considered as invariant, but it the process were repeated then the boundary could be at a different phase with respect to the source. The effects can be assembled as receiver amplitude versus iteration number (time) plots where the time refers to the motion of a sea wave. By this means we could investigate the effects of surface waves on the response at each of the two receivers.

Further extensions of these ideas could include investigations of:

(a) the case where the wave-length of the surface boundary is of the same order as the wavelength of the acoustic signal;

(b) trochoidal wave profiles

$$x = A[3\cos\theta + \cos 3\theta]$$
$$y = A[3\sin\theta - \sin 3\theta] \ (0 < \theta < \pi).$$

6.4 In the chapter on rule-based models Fig. 5.5 related to the dispersion of smoke from a chimney in the presence of a wind. The associated scattering matrix was given by:

$$
\begin{pmatrix} {}^sQ_1 \\ {}^sQ_2 \\ {}^sQ_3 \\ {}^sQ_4 \\ {}^sQ_5 \\ {}^sQ_6 \end{pmatrix}_k = \frac{1}{6}
\begin{pmatrix}
1 & 1 & 1 & 1 & 1 & 1 \\
1 & 1-6\alpha & 1 & 1-6\alpha & 1 & 1 \\
1 & 1 & 1 & 1 & 1 & 1 \\
1 & 1+6\alpha & 1 & 1+6\alpha & 1 & 1 \\
1 & 1 & 1 & 1 & 1 & 1 \\
1 & 1 & 1 & 1 & 1 & 1
\end{pmatrix}
\begin{pmatrix} {}^iQ_1 \\ {}^iQ_2 \\ {}^iQ_3 \\ {}^iQ_4 \\ {}^iQ_5 \\ {}^iQ_6 \end{pmatrix}_k .
$$

This equation should be adapted to allow for a randomly fluctuating bias, α, which might express local variations of the wind within a single experiment. This could account for factors such as eddy effects.

Fig. 6.17 Chimney-smoke ellipse.

If the behaviour of the wind over time can be described by an ellipsoidal plot with a major to minor diameter of 3:1 (see Fig. 6.17), then develop a model that reflects the changes in direction and magnitude of α in a statistical way. This could then be used to determine the geographical region of maximum smoke concentration in relation to the chimney over a prolonged timescale.

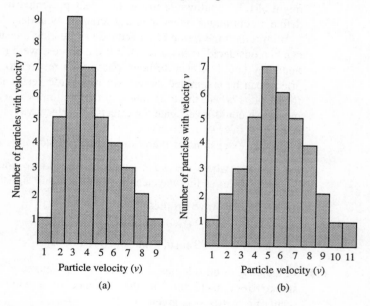

Fig. 6.18 Distributions of particle velocities.

6.5 Construct a Monte Carlo model consisting of 100 particles which can assume velocities according to distribution of Fig. 6.18(a). In the first instance the problem space will consist of 100×100 nodes. Initially the particles will be randomly distributed in space and will have random velocities within the distribution and will travel in any one of four directions. A particle which had velocity '9' would then move 9 nodes in that direction, unless it encountered another particle *en route*. If a collision occurs, then the dynamics depends on the relative velocities.

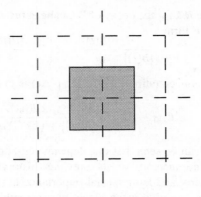

Fig. 6.19 Cell on a Cartesian lattice.

Collisions with the boundaries of the mesh are elastic. For the purpose of this experiment a region of boundary consisting of 10 nodes is monitored and the number of particles which hit this region within 100 time steps is monitored. This process can be done several times in order to ensure that equilibrium has been established and the rate of arrival is no longer dependent on the initial distributions.

Boyle's law can be simulated by repeating the simulation for the same number of particles, but now occupying a smaller problem space (e.g. 90 nodes × 90 nodes). The dependence of mean rate of arrival within the monitored region (pressure) as a function of mesh area (volume) can then be plotted.

Distribution (a) of Fig. 6.18 was intended to be a rough approximation to a Maxwell–Boltzman distribution at a low temperature. The effect of increasing the mean energy of particles (temperature) can be simulated by repeating the exercise using the distribution of Fig. 6.18 (b).

The model can then be moved from an ideal gas towards a real gas. Let us assume that the forces of attraction are small, because we have a large volume (100 nodes × 100 nodes), but let us assume that and particle occupies a cell as shown in Fig. 6.19. If any other particle comes to within one unit distance in either the *x* or *y* directions then a collision will occur. This effectively increases the probability of a collision.

Consider how you might introduce an attraction between particles and an attraction between particles and the boundaries of the problem. One approach might be to increase the velocity of a particle in a particular direction at the next iteration which is dependent on the separation between that particle and another particle or the surface.

6.6 The instantaneous concentration of a species in a birth/death chemical reaction of the form A → B → C with rate *R* for the generation of B and a rate R/k for the decay of B can be expressed by the differential equation of the form

$$\frac{1}{[B(t)]} \frac{d[B(t)]}{dt} = R\left\{1 - \frac{[B(t)]}{k}\right\}.$$

The concentration of B at $(t_o + \Delta t)$ is given by

$$[B(t_0 + \Delta t)] = \frac{k[B(t_0)]}{[B(t_0)] + (k - B[t_0])e^{-R\Delta t}}.$$

It can be seen that this depends only on the concentration at t_o, i.e. it has no memory of its previous history. This is called the Markov property and is of special importance in probabilistic modelling.

An operative in an electronics assembly factory 'reworks' defective printed circuit boards at the rate of R per hour. If a board arrives on the bench at $t = 0$ then we can look at the following discrete probabilities:

$r_{00}(t)$: probability that the job is finished at t if it is finished at t_o

$r_{01}(t)$: probability that the job is unfinished at t if it is finished at t_o

$r_{10}(t)$: probability that the job is finished at t if it is unfinished at t_o

$r_{11}(t)$: probability that the job is unfinished at t if it is unfinished at t_o.

These options can be assembled into a stochastic matrix where the sum of any row is unity

$$[R(t)] = \begin{bmatrix} r_{00}(t) & r_{01}(t) \\ r_{10}(t) & r_{11}(t) \end{bmatrix}.$$

Obviously $r_{00}(t) = 1$ and $r_{01}(t) = 0$ so that we can write

$$[R(t)] = \begin{bmatrix} 1 & 0 \\ 1 - r_{11}(t) & r_{11}(t) \end{bmatrix}.$$

It the probability of a board being processed during a short time dt is Rdt, then show that in the limit the probability of a board being unfinished at t, given that it was unfinished at t_0 is given by

$$r_{11}(t) = e^{-Rt}.$$

The concept can be extended. If there were N boards to be processed then the stochastic matrix is lower triangular with $N+1$ rows and $N+1$ columns. This is shown below for $N = 4$.

$$[R(t)] = \begin{bmatrix} 1 & 0 & 0 & 0 & 0 \\ r_{10}(t) & e^{-Rt} & 0 & 0 & 0 \\ r_{20}(t) & Rte^{-Rt} & e^{-Rt} & 0 & 0 \\ r_{30}(t) & \frac{1}{2}(Rt)^2 e^{-Rt} & Rte^{-Rt} & e^{-Rt} & 0 \\ r_{40}(t) & \frac{1}{3!}(Rt)^3 e^{-Rt} & \frac{1}{2}(Rt)^2 e^{-Rt} & Rte^{-Rt} & e^{-Rt} \end{bmatrix}$$

where

$$r_{N0}(t) = 1 - e^{-Rt} \sum_{k=0}^{N-1} \frac{(Rt)^k}{k!}.$$

This last expression gives the probability that N boards will be processed at t given that none were processed at the start.

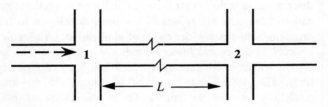

Fig. 6.20 Road layout for Problem 6.7.

6.7 The diagram in Fig. 6.20 shows two intersections on a road which are controlled by traffic lights.

Use a GPSS-type approach (Section 6.3.2) to investigate the influence of duty cycle ($t_{go}/(t_{go} + t_{stop})$) of lights 1 and 2 and their relative phasing on traffic flow patterns in the direction indicated by the arrow as a function of separation between the two intersections.

Non-Cartesian meshes have been encountered elsewhere in this book. The cylindrical problems of Sections 2.2.3, 2.4.2, and 2.4.3 could be treated using either the finite-difference or transmission-line-matrix methods. However, we intend to cover more general situations here; cases where the mesh consists of curvilinear rectangles which follow geometrical boundaries or where the discretization is no longer rectangular. The first section introduces a technique called *tubes and slices* which has been developed to treat electromagnetic problems, but has a wider range of application. Of specific interest here is the concept of upper and lower bounds for a solution. The second section deals with a widely-used technique which has been extensively covered in many texts. The *finite-element method* is unusual in the area of mathematical modelling in that its origins lie in the fields of mechanical and civil engineering. Because of the level of mathematical complexity which is involved in all but the simplest problems it is not normal for the user to develop *ab initio* finite-element code. There are many commercially available packages, most of which are tailored to specific applications. The treatment here is therefore restricted to an outline of the method.

7.1 Tubes and slices

The concept of *tubes and slices* was introduced by Hammond [7.1]. It has been further developed in a companion volume to this text [7.2] which has accompanying software. The method operates on the basis that flow of electricity in a conductor of arbitrary shape is determined by resistance. The region between the two contacts shown in Fig. 7.1(a) can be represented by many smaller resistors connected in series or by many larger resistors connected in parallel (Fig. 7.1(b)). The total resistance of the material as shown in Fig. 7.1(a) is given by $R = \Sigma R_s$, while the same using Fig. 7.1(b) can be obtained from $R^{-1} = \Sigma (R_p)^{-1}$.

The same representations could be obtained in the following way. Let us suppose that the material in Fig. 7.1(a) was broken into n segments with each segment joined to the next by a perfect conductor so that the overall resistance was not affected. The resistance of each segment could be given as:

$$R_s = \frac{L/n}{\sigma A} \tag{7.1}$$

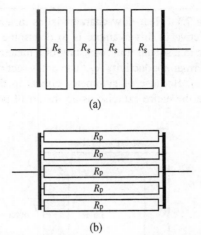

(a)

(b)

Fig. 7.1 (a) A conductor represented as a series connection of resistors. (b) A similar representation using parallel connected resistors.

where L is the sample length, A is the sample area, and σ is the electrical conductivity. Thus $R = nR_s$.

Similarly, if the discretization of Fig. 7.1(b) was broken into m segments where each segment was separated by a piece of perfect insulator then the overall resistance would be unaffected by this change and would be given as:

$$R_p = \frac{L}{\sigma A/m}. \tag{7.2}$$

Thus $R = R_p/m$.

Because of the physical representation the parallel resistors connecting two electrodes are called *tubes*, while the series resistors are called *slices*. Hammond and Sykulski [7.2] highlight an argument put forward by Maxwell. He reasoned that if a conductor was represented by tubes only then the insertion of thin insulators would have no effect on the resistance provided that they followed the contours of the current flow. If they did not, then the effect would be to raise the resistance above its unperturbed value. Similarly, the perfect conductors placed between slices would leave the resistance undisturbed so long as they lay along lines of equal potential. If not, then the resistance would be less. In other words, the perfect arrangement of tubes gives a minimum value (lower bound) for resistance while the perfect arrangement of slices give a maximum or upper bound. The method of obtaining an approximation of the exact result consists of taking the average between the upper bound (R_+) and the lower bound (R_-). The capacitance between two electrodes can also be obtained within upper and lower bounds. The accuracy of the estimate is very good in both cases.

Example 7.1 Heat-flow characteristics in a rectangular duct

The objective in this example is to determine the thermal capacitance and resistance in a square flue consisting of a material of specific heat C_p, density ρ and thermal conductivity k_T. The cross-sectional dimensions are shown in Fig. 7.2(a). Because of symmetry effects in this structure it is possible to undertake the entire exercise using the small portion shown in Fig. 7.2(b).

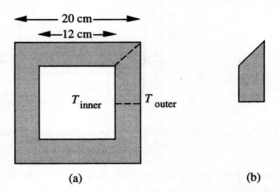

(a)   (b)

Fig. 7.2 (a) A square duct. (b) The study region.

The segment is shown again in Fig. 7.3 (a) and (b) where it is broken into slices and tubes. The sum of these gives

$$R_s = \frac{1}{k_T}\left[\frac{1}{6.5} + \frac{1}{7.5} + \frac{1}{8.5} + \frac{1}{9.5}\right] = \frac{0.51}{k_T}$$

as a lower bound.

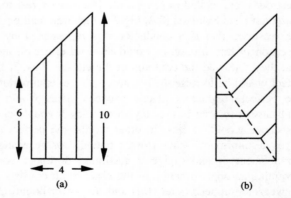

(a)   (b)

Fig. 7.3 The slice (a) and tube (b) discretizations of the portion of the square duct of Fig. 7.2.

The tube discretization involves further segmentation on either side of the dashed line shown in Fig. 7.3(b). Thus any tube is the sum of two components in series. The resistance of the lowermost tube is given by:

length $(4+3)/2$ divided by k_T times area $1 \times 6/4$ for the first part plus length $(\sqrt{4^2 + 4^2})/(2 \times 4)$ divided by k_T times area $1 \times \sqrt{6^2 + 4^2}/4$. That is

$$R_1 = \frac{3.5}{k_T(1.5)} + \frac{\left(\dfrac{\sqrt{32}}{8}\right)}{k_T(2.06)} = \frac{2.676}{k_T}.$$

The area, 2.06 is derived from $\sqrt{(10/4)^2 - (4^2 + 4^2)/4^2}$. Similarly

$$R_2 = \frac{2.5}{k_T(1.5)} + \frac{\left(3\dfrac{\sqrt{32}}{8}\right)}{k_T(2.06)} = \frac{2.695}{k_T}$$

$$R_3 = \frac{1.5}{k_T(1.5)} + \frac{\left(5\dfrac{\sqrt{32}}{8}\right)}{k_T(2.06)} = \frac{2.715}{k_T}$$

$$R_4 = \frac{0.5}{k_T(1.5)} + \frac{\left(7\dfrac{\sqrt{32}}{8}\right)}{k_T(2.06)} = \frac{2.734}{k_T}.$$

The total resistance of the tubes is then given by the sum of the reciprocals

$$\frac{1}{R_T} = k_T\left[\frac{1}{2.676} + \frac{1}{2.695} + \frac{1}{2.715} + \frac{1}{734}\right] = 1.478\,k_T$$

Thus giving

$$R_T = \frac{0.676}{k_T}$$

which is an upper bound. The thermal resistance of the segment is then taken as the mean of R_T and R_S which is $0.593/k_T$. Since this is one eighth of the entire flue the total resistance is $0.074/k_T$ per unit depth. When the same problem is analysed using the *Tubes and Slices* package in ref. [7.2] the result is also $0.074/k_T$, although the authors use a more sophisticated subdivision as shown in outline in Fig. 7.4.

Exactly the same geometry could be applied to a problem in electrostatics to determine capacitance on the basis that tubes are equivalent to capacitors in parallel and yield $C_T = C_1 + C_2 + \ldots$ (an upper bound), while slices are capacitors in series and give $1/C_S = 1/C_1 + 1/C_2 + \ldots$ (a

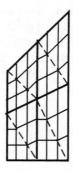

Fig. 7.4 A tube-and-slice discretization where each of the boundary lines has been bisected to give four quadrilaterals. The midpoints are joined by diagonals (broken lines) for construction purposes. The slices are uniform and within each triangle the tubes are parallel to one side.

lower bound). However, it must be remembered that the normal electrical analogue of thermal capacitance is a capacitor shunted to ground (see Fig. 4.17). The physical interpretation of this is that all the capacitors are effectively in parallel. For instance, we can take the slices of Fig. 7.3(a) and use the equation for thermal capacitance (ρC_p volume) to obtain.

$$C_S = \rho C_p \, (6.5 + 7.5 + 8.5 + 9.5) = 32\rho C_p.$$

As this segment is one eighth of the entire duct it can be seen that the total capacitance based on slices is $256\rho C_p$ per unit length of flue, which is identical with the value calculated on the basis of $\rho C_p(20^2 - 12^2)$. It is left to the reader to confirm that a calculation using tubes would give the same answer.

The tubes-and-slices method gives very reasonable results, even in cases such as triangular meshes where Cartesian coordinates would appear to be the most obvious method of description. It also demonstrates the concept of upper and lower bounds of accuracy defined by the series and parallel components of capacitance. Triangular meshes are of particular importance in finite-element modelling, which we shall consider next.

7.2 Finite elements

The individual steps involved in the finite-element technique are simple enough, but when put together they result in a method which can be mathematically intensive and which tends to divorce the modeller from the problem being considered. That said, it is an important technique, and the

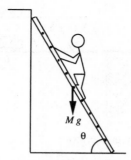

Fig. 7.5 Force acting at the centre of gravity of a ladder placed against a wall. The mass M represents the sum of the masses of the ladder and the user.

reader is referred to some of the specialist texts [e.g. 7.3, 7.4]. The method can be described as being concerned with forces, work, and equilibrium. It largely evolved from problems in mechanics and was in the first instance widely applied in civil engineering.

7.2.1 Finite element modelling of discrete systems

To illustrate the technique, we start by considering a ladder placed against a wall at an angle θ (Fig. 7.5) forming a triangle. The angle θ is chosen to ensure stability of the ladder when unloaded. As a potential user, we may wish to know whether the ladder will still be stable if we climb it. Obviously this is critically dependent upon the angle θ. The force acting through the centre of gravity of the ladder (Fig. 7.5) is resolved as a force acting on the wall and a force acting on the ground. If the horizontal component of force on the ground and the vertical component on the wall exceed the opposing frictional forces then the ladder will slide in order to minimize θ!

At a more detailed level we might be interested in estimating the effects of loading the ladder with a particularly heavy user. Obviously the ladder will bend slightly and the contact positions on the wall and ground will be moved by a small amount. We know that work is defined as Fx (force times displacement). If we accept the principle that the total change in potential energy (work done) is always the minimum neccessary for the task then it should be possible to caculate the actual displacements for a given length of ladder, loading, and angle of inclination.

On a larger scale we might consider the effect of a heavy vehicle on the mechanical stability of a road bridge (Fig. 7.6). This bridge design is typical of many and the structure support consists of a triangular arrangement of steel girders. This resembles the type of two-dimensional spatial discretization which is familiar to finite-element (FE) modellers. A civil engineer might be interested in the stresses on and displacement of these structural members.

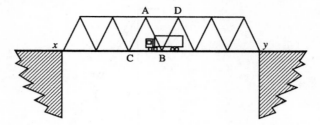

Fig. 7.6 Loading on a bridge.

Looking at the member CB it is obvious that it will be stretched by the weight of the bridge itself, and will be stretched even more as the lorry passes over. At the same time member AD will be compressed. If the stress on CB causes it to stretch beyond its elastic limit there will be permanent deformation of the bridge. If the fracture limit of material is exceeded then the bridge is in danger of collapse. Similarly, AD will have a buckling limit. The bank seats (support points x and y) may be steel or may be concrete and if the latter there will be a crushing limit.

This problem could be analysed using an electrical analogue. The approach might be to obtain an expression for one element (triangle) and then extend the treatment to as many as are required. We can demonstrate this by means of several examples which will move towards the problem in Fig. 7.6. The first of these outlined below is equivalent to a set of resistors connected in series.

Example 7.2 The thermal characteristics of a furnace
A conventional high temperature furnace consists of an operating cavity which is surrounded by a lining of fire bricks which have low thermal conductivity and are resistant to the high temperatures. This is surrounded by an outer lining which may comprise the majority of the volume of the furnace. The outer casing may be sheet steel or some other material. This is shown in schematic form in Fig. 7.7. The finite-element method will be used to calculate the unknown temperatures T_2 and T_3 (at interfaces 2 and 3 respectively) as well as the heat flow per unit area.

On the basis of unit area we can develop a series of matrix equations relating heat flow and temperature. Considering the fire brick, we can write two heat-flow equations for this 'element':

$$\frac{k_1}{L_1}(T_1 - T_2) = Q_1$$

$$\frac{k_1}{L_1}(T_2 - T_1) = Q_2.$$

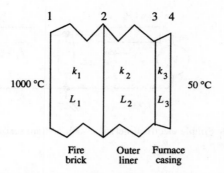

Fig. 7.7 Schematic cross section of a furnace showing the thermal conductivities (k) and widths (L) of the components.

This of course ignores any equality of heat flow in an equilibrium situation. These can be expressed in matrix form as:

$$\frac{k_1}{L_1}\begin{bmatrix} 1 & -1 \\ -1 & 1 \end{bmatrix}\begin{bmatrix} T_1 \\ T_2 \end{bmatrix} = \begin{bmatrix} Q_1 \\ Q_2 \end{bmatrix}.$$

There are similar expressions for heat flow in the other two elements and using the analogy k/L is thermal conductance (K) per unit area we can assemble all of these in a connectivity matrix

$$\begin{bmatrix} K_1 & -K_1 & 0 & 0 \\ -K_1 & K_1 + K_2 & -K_2 & 0 \\ 0 & -K_2 & K_2 + K_3 & -K_3 \\ 0 & 0 & -K_3 & K_3 \end{bmatrix}\begin{bmatrix} T_1 \\ T_2 \\ T_3 \\ T_4 \end{bmatrix} = \begin{bmatrix} Q_1 \\ 0 \\ 0 \\ Q_4 \end{bmatrix}.$$

Here T_1 and T_4 are the given internal and external temperatures, which leaves four unknowns, T_2, T_3, Q_1, and Q_4 to be solved. Of course, at equilibrium there is a heat-flow balance, $Q_1 = Q_4$. The connectivity matrix of this example should not be a problem to students in electronic/electrical engineering as it merely expresses network connectivity as the next example demonstrates.

Example 7.3 Finite-element solution of some simple electical networks
In electrical terms the previous example was equivalent to three resistors in series. The example shown in Fig. 7.8 has a shunt resistance and is intended to reinforce the ideas of the connectivity matrix.

Students of electronic/electrical engineering will again recognize that this circuit can be analysed using any one of several well-known methods which are based on Kirchoff's laws. Those who are less familiar with this

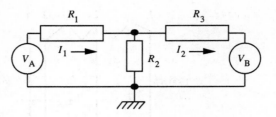

Fig. 7.8 Simple electrical network with a shunt resistor.

approach might like to consult introductory texts such as Howatson [7.5]. The two potentials can be equated as:

$$V_A = I_1 R_1 + I_1 R_2 - I_2 R_2$$
$$-V_B = I_2 R_2 + I_2 R_3 - I_1 R_2$$

Which can be expressed in matrix terms as:

$$\begin{bmatrix} R_1 + R_2 & -R_2 \\ -R_2 & R_2 + R_3 \end{bmatrix} \begin{bmatrix} I_1 \\ I_2 \end{bmatrix} = \begin{bmatrix} V_A \\ -V_B \end{bmatrix}.$$

If V_A and V_B are known then it is possible to determine the value of I_1, I_2.

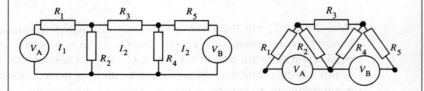

Fig. 7.9 (a) A three unit ladder electrical network. (b) An alternative representation of the same ladder network.

A ladder network (Fig. 7.9(a)) can be formed by joining together two of the units of Fig. 7.8 and the matrix statement of Ohm's law then reads:

$$\begin{bmatrix} R_1 + R_2 & -R_2 & 0 \\ -R_2 & R_2 + R_3 + R_4 & -R_4 \\ 0 & -R_4 & R_4 + R_5 \end{bmatrix} \begin{bmatrix} I_1 \\ I_2 \\ I_3 \end{bmatrix} = \begin{bmatrix} V_A \\ 0 \\ -V_B \end{bmatrix}$$

which can be seen as the merging of the units in matrix terms. Once again, provided we have enough information, it is possible to determine the values of the unknown parameters. Inspection of Fig. 7.9(b) shows how a simple re-arrangement of this circuit can provide an interesting analogue of part of the bridge structure of Fig. 7.6.

Thus it should be possible to analyse this bridge or a structure such as an overhead electrical pylon using such an approach. The equations can be decribed in terms of simple units and the connectivity of the problem is reflected in the topology of the resistance matrix.

7.2.2 Finite element modelling of continuous systems

We might now ask whether it is possible to undertake a similar analysis on a continuous medium such as a concrete bridge or a sheet-steel chimney. The answer is yes. The principles are similar but the practice is somewhat more complicated. There are essentially two different approaches: the energy minimization method and the residual method. The former will be presented in terms of an example. An introduction to the residual FE approach will be preceded by outline treatments of two essential components: spatial discretization and the use of residual methods to solve differential equations.

FE modelling using energy methods

This approach involves the following steps:

- subdivision of the problem space into small elements;

- development of a description for the unknown parameters inside each element in terms of the nodal values;

- development of a matrix which describes the response of an element to externally applied forces, remembering that changes in potential energy will be minimized;

- assembly of the elements into a connectivity matrix which describes the shape and boundaries of the physical problem;

- solution of the system of equations using matrix methods.

Example 7.4 Electrical characteristics of buried conductors
This example, which is due to Silvester and Ferrari [7.3], goes one step further than the previous ones in an attempt to demonstrate the way in which the finite element technique can be extended from discrete elements to elements consisting of a continuous medium (the question raised in

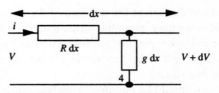

Fig. 7.10 The resistance and conductance in a small element (length dx) consisting of two parallel conductors buried in a conductive medium.

connection with Fig. 7.6). It concerns two parallel (uninsulated) cables which are buried some distance apart in the ground. This means that in addition to the resistance of the cables themselves, there will also be some conductance between the cables. If we consider an element of length dx we can represent the resistance and conductance as shown in Fig. 7.10.

Silvester's first step is to divide the length into a series of elements numbered 1, 2, 3, ... The potentials at the left and right terminals of an element are designated V_1 and V_r. On the basis that the voltage varies linearly along an element we can develop an expression for the voltage at position x within the jth element:

$$V = \frac{x(j)_r - x}{x(j)_r - x(j)_1} V(j)_1 + \frac{x - x(j)_1}{x(j)_r - x(j)_1} V(j)_r.$$

or in short form: $V = \alpha(j)_1 V(j)_1 + \alpha(j)_r V(j)_r$

where $\alpha(j)_1 = \dfrac{x(j)_r - x}{x(j)_r - x(j)_1}$ and $\alpha(j)_r = \dfrac{x - x(j)_1}{x(j)_r - x(j)_1}$.

The total power dissipated in the system is the sum of the powers lost in each element. The power lost in any element is

$$W(j) = - \int\limits_{x(j)_1}^{x(j)_r} \left[g\, V^2 + \frac{1}{r} \left(\frac{dv}{dx} \right)^2 \right] dx$$

which after substitution of the linearly dependent voltage yields

$$W(j) = -\frac{1}{r(j)} \int\limits_{x(j)_1}^{x(j)_r} \left[V(j)_1 \frac{d\alpha(j)_1}{dx} + V(j)_r \frac{d\alpha(j)_r}{dx} \right]^2 dx$$

$$-g(j) \int\limits_{x(j)_1}^{x(j)_r} [\alpha(j)_1 V(j)_1 + \alpha(j)_r V(j)_r]^2 dx.$$

In order to simplify the mathematics we can assume that the parameters r and g are constant within any element. The power lost within an element can then be expressed in matrix form as:

$$W(j) = -\left[V(j)_1\ V(j)_r\right]\left[\frac{1}{r(j)}S + g(j)T\right]\left[\begin{array}{c}V(j)_1\\V(j)_r\end{array}\right].$$

The matrices S and T have terms

$$S_{ij} = \int\limits_{x_1}^{x_r} \frac{d\alpha_i}{dx}\frac{d\alpha_j}{dx}\ dx \text{ and } T_{ij} = \int\limits_{x_1}^{x_r} \alpha_i\ \alpha_j\ dx$$

with the matrix subscripts i and j having values 1 and r.

This can be compacted further since

$$\frac{1}{r(j)}S + g(j)\ T = M$$

and then, using $V(j)$ and $V^T(j)$ as the vector and transpose of the left- and right-end voltages of the jth element, we finally have the following expression for power:

$$W(j) = V^T(j)\ M\ V(j).$$

Silvester and Ferrari [7.3] discuss how the approach can be modified to treat each element in terms of normalized variables so that the matrix can be applied generally (i.e. it is independent of any particular element).

Having developed a set of element matrices, the next stage is to produce a matrix which describes the connection of the individual elements. In terms of electrical/electronic engineering this problem can be considered as a cascade of two-port networks and there is a matrix equation which expresses the relationship between the disconnected voltages V_{dis} and the connected voltages V_{con}:

$$V_{\text{dis}} = CV_{\text{con}}$$

(C is the connection matrix). Using this equation we can move from the work done by an individual element to the work done by the entire system. The previous expression for power was in terms of disconnected elements

$$W = V_{\text{dis}}^T\ M_{\text{dis}}V_{\text{dis}}$$

and so

$$W = V_{\text{con}}^T C^T M_{\text{dis}}C\ V_{\text{con}}$$

which suggest a transformation

$$M = C^T M_{\text{dis}} C.$$

On the basis of five elements of equal length, L, it can be shown that

$$\frac{1}{r} C^T S C = \frac{1}{L\,r} \begin{bmatrix} 1 & -1 & 0 & 0 & 0 & 0 \\ -1 & -2 & -1 & 0 & 0 & 0 \\ 0 & -1 & 2 & -1 & 0 & 0 \\ 0 & 0 & -1 & 2 & -1 & 0 \\ 0 & 0 & 0 & -1 & 2 & -1 \\ 0 & 0 & 0 & 0 & -1 & 1 \end{bmatrix}$$

and

$$g C^T T C = \frac{L\,g}{6} \begin{bmatrix} 2 & 1 & 0 & 0 & 0 & 0 \\ 1 & 4 & 1 & 0 & 0 & 0 \\ 0 & 1 & 4 & 1 & 0 & 0 \\ 0 & 0 & 1 & 4 & 1 & 0 \\ 0 & 0 & 0 & 1 & 4 & 1 \\ 0 & 0 & 0 & 0 & 1 & 2 \end{bmatrix}.$$

The final step in the entire analysis is to minimize the total power in terms of each of the nodal voltages. In the problem under consideration every node is free to vary except the value at the source ($j = 1$). Thus we can differentiate the expression with respect to each $V(j)$ which is treated as a variable.

$$\frac{\partial W}{\partial V(j)} = 0 \ (j = 1 \text{ to } N - 1).$$

We obtain a matrix of N columns (one for each node) and N–1 rows (one for each of the voltages that is free to vary) which can be solved to obtain the value of the voltages $V(j)$ at each node.

The element shape function

The previous example dealt with a continuous problem in one dimension. In the case of a two-dimensional problem we need to be in a position to undertake our calculations at fixed positions and yet be able to describe the properties at any point within a region (element). For that reason we must consider the element shape function.

We can assume that the potential at any point within an element in a two-dimensional electrostatics problem can be described by a general expression as:

$$\tilde{V}^e = a + bx + cy + dx^2 + exy + fy^2 + \ldots . \tag{7.3}$$

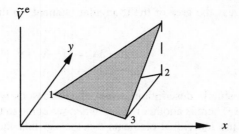

Fig. 7.11 A trial solution represented by a triangular element shaded and its projection on the $x - y$ plane.

For reasons which will become obvious later this is called a trial solution and comprises a constant term, two linear terms, three quadratic terms, and so on. In many cases an adequate approximation can be obtained by considering only the linear terms of the polynomial and this will be considered first.

If we assume that the potential at any point within the triangular element shown in Fig. 7.11 can be described by a linear approximation then

$$\tilde{V}^e = a + bx + cy. \tag{7.4}$$

Now, the nodal ('known') potentials can be expressed in matrix form as:

$$\begin{bmatrix} V_1 \\ V_2 \\ V_3 \end{bmatrix} = \begin{bmatrix} 1 & x_1 & y_1 \\ 1 & x_2 & y_2 \\ 1 & x_3 & y_3 \end{bmatrix} \begin{bmatrix} a \\ b \\ c \end{bmatrix} \tag{7.5}$$

so that a, b, and c can be determined.

If the values of a, b, and c from eqn (7.5) are inserted into eqn (7.4) then the espression for $\tilde{V}^e$ can be written as:

$$[1 \ x \ y]\frac{1}{2A} \begin{bmatrix} x_2y_3 - x_3y_2 & x_3y_1 - x_1y_3 & x_1y_2 - x_2y_1 \\ y_2 - y_3 & y_3 - y_1 & y_1 - y_2 \\ x_3 - x_2 & x_1 - x_3 & x_2 - x_1 \end{bmatrix} \begin{bmatrix} V_1 \\ V_2 \\ V_3 \end{bmatrix} = \sum_{i=1}^{3} \alpha_i(x, y) V_i \tag{7.6}$$

where the element shape functions are given by

$$\alpha_1 = \frac{1}{2A}[(x_2y_3 - x_3y_2) \quad (y_2 - y_3)x \quad (x_3 - x_2)y]$$

$$\alpha_2 = \frac{1}{2A}[(x_3y_1 - x_1y_3) \quad (y_3 - y_1)x \quad (x_1 - x_3)y] \tag{7.7}$$

$$\alpha_3 = \frac{1}{2A}[(x_1y_2 - x_2y_1) \quad (y_1 - y_2)x \quad (x_2 - x_1)y]$$

and A is the area of the triangular element, so that

$$2A = \begin{vmatrix} 1 & x_1 & y_1 \\ 1 & x_2 & y_2 \\ 1 & x_3 & y_3 \end{vmatrix}.$$

This simple description is one of the reasons why triangles are favoured in FEM. There is another reason: when we come to the next stage we will need to join the individual elements together so that the region of study is covered continuously by *connected* elements. It is a feature of triangular discretizations that if continuity is prescribed at the common nodes of two joined triangles, then continuity is guaranteed all along the common boundary. This means that if the two elements are joined together along an edge, they will have two common apexes, nodes which have the same potential. If this is the case, then for a triangle, the potentials will be common all along the shared edge.

Spatial discretization and triangulation

Having indicated the importance of triangular elements in FEM, it might be worth some discussion about methods of discretizing a two-dimensional spatial region.

(a) Rectangular domains

We can divide any area into a series of small rectangles and we can also have mesh refinement so that significant areas have a smaller discretization. Diagonalization of the rectangles yields a triangular discretization. In general, if we have n_x and n_y divisions then the number of elements $= 2n_x n_y$ and the number of nodes $= (n_x + 1)(n_y + 1)$. This is shown in Fig. 7.12.

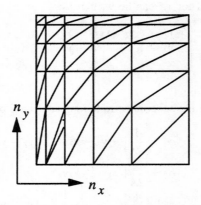

Fig. 7.12 Graded discretization of a rectangular domain.

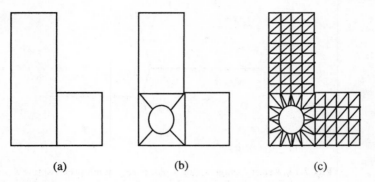

Fig. 7.13 Discretizations of an arbitrary domain with and without a hole.

(b) Arbitrary domains with segmentation
The entire region of Fig. 7.13(a) has been divided into two quadrilateral blocks. Each block can then be treated as in (a) and then rejoined. If an arbitrary domain has a feature such as a hole (Fig. 7.13(b)) then the situation becomes slightly more difficult. A usable discretization is shown in Fig. 7.13(c).

(c) Arbitrary domains by Delaunay triangulation
At an early stage in the development of FEM discretizations it was realized that a higher level of accuracy in the approximations could be obtained if the triangular discretization process managed to avoid the use of extreme values of angles. One method of achieving a discretization that attempts to maximize all angles is Delaunay triangulation, which is well known in land surveying. There is a *dual relationship* between this and Voronoi diagrams. If a set of *n* points are chosen within an arbitrary region then the lines which are equidistant from two or more points are called Voronoi edges and the intersection between two edges defines a Voronoi vertex. If we could visualize a set of fire observation towers in a forest with each having a region of responsibility, then this might be as shown in Fig. 7.14.

Now, if we have the Voronoi diagram for a set of 6 points within a bounded region and if lines are then drawn so as to cross Voronoi edges at right angles, then the result, shown in bold-lines in Fig. 7.15 is a Delaunay triangulation.

The choice of triangles can be made using the *circle test*. Figure 7.16 shows a polygon which is to be triangulated by means of a diagonal. The attempt on the left-hand side produces two triangles. If a circumcircle is drawn around triangle 1 it will be seen that the apex of triangle 2 falls within the same circle. This is *not* a valid Delauny triangulation. The same result would have been obtained had the circumcircle been drawn around triangle 2. At this point the diagonal line is swapped to yield triangles 1' and 2' as shown on the right. If a

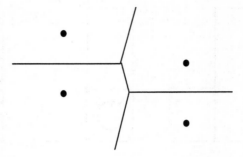

Fig. 7.14 Voronoi diagram for 4 points (e.g. five observation towers in a forest, each with a clearly defined non-overlapping region of responsibility).

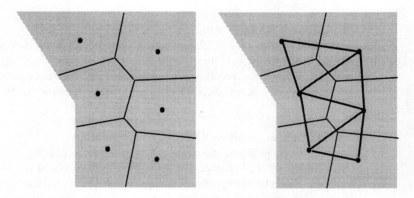

Fig. 7.15 (a) A 6 point Voronoi diagram. (b) Same diagram with Delaunay triangulation superimposed.

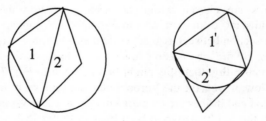

Fig. 7.16 Circle tests for the triangulation of a polygon.

circumcircle is now drawn around one of these triangles then it will be seen that the apex of the other falls outside the circle. This is therefore a valid Delaunay triangulation.

Armed with information on how to discretize a two-dimensional medium we can see how this is applied in a simple Laplace-type example.

Example 7.5 Equilibrium temperature in quadrilateral section of material
Figure 7.17 shows a section of material which has been discretized into two triangular elements. There is much information present. The elements are identified by bold numbering and the dimensions and boundary conditions are also shown.

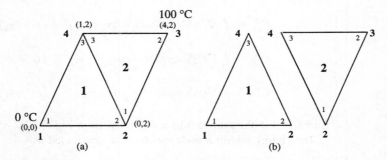

Fig. 7.17 Quadrilateral divided into two elements numbered in large bold letters. The boundary values of temperature are also given. The coordinates are shown in parentheses. The element and global numbering is shown for (a) the connected system and (b) the discretized system.

The functional corresponding to Laplace's equation can be written in terms of an electrical analogue as:

$$W^e = \frac{1}{2} \int \varepsilon |\nabla \tilde{V}^e|^2 dS$$

where

$$\nabla \tilde{V}^e = \sum_{i=1}^{3} V_i \nabla \alpha_i$$

so that

$$W^e = \frac{1}{2} \varepsilon \sum_{i=1}^{3} \sum_{j=1}^{3} V_i \left[\int \nabla \alpha_i \nabla \alpha_j \, dS \right] V_j.$$

The term inside the square brackets can be designated by C_{ij}^e and can be derived as follows:

$$C_{12}^e = \int \nabla \alpha_1 \nabla \alpha_2 \, dS = \frac{1}{4A^2} [(y_2 - y_3)(y_3 - y_1) + (x_3 - x_2)(x_1 - x_3)] \int dS$$

$$= \frac{1}{4A} [(y_2 - y_3)(y_3 - y_1) + (x_3 - x_2)(x_1 - x_3)] = C_{21}^e$$

$$C_{13}^e = \int \nabla\alpha_1 \nabla\alpha_3 \, dS = \frac{1}{4A} \left[(y_2 - y_3)(y_1 - y_2) + (x_3 - x_2)(x_2 - x_1) \right] = C_{31}^e$$

$$C_{23}^e = \int \nabla\alpha_2 \nabla\alpha_3 \, dS = \frac{1}{4A} \left[(y_3 - y_1)(y_1 - y_2) + (x_1 - x_3)(x_2 - x_1) \right] = C_{32}^e.$$

And the diagonal terms are:

$$C_{11}^e = \int \nabla\alpha_1 \nabla\alpha_1 \, dS = \frac{1}{4A} \left[(y_2 - y_3)^2 + (x_3 - x_2)^2 \right]$$

$$C_{22}^e = \int \nabla\alpha_2 \nabla\alpha_2 \, dS = \frac{1}{4A} \left[(y_3 - y_1)^2 + (x_1 - x_3)^2 \right]$$

$$C_{33}^e = \int \nabla\alpha_3 \nabla\alpha_3 \, dS = \frac{1}{4A} \left[(y_1 - y_2)^2 + (x_2 - x_1)^2 \right]$$

(in each case the numberings are derived from Fig. 7.17).

The energy within a node can now be designated as:

$$W^e = \frac{1}{2}\varepsilon [\tilde{V}^e]^T [C^e] [\tilde{V}^e]$$

where $[\tilde{V}^e]^T$ is the transpose of $[\tilde{V}^e]$ and $[C^e]$ is the element coefficient or stiffness matrix given by:

$$[C^e] = \begin{bmatrix} C_{11}^e & C_{12}^e & C_{13}^e \\ C_{21}^e & C_{22}^e & C_{23}^e \\ C_{31}^e & C_{32}^e & C_{33}^e \end{bmatrix}.$$

The coordinates shown in Fig. 7.17 can be now used to obtain a stiffness matrix for each element in this problem, $[C^{(1)}]$ and $[C^{(2)}]$:

$$[C^{(1)}] = \begin{bmatrix} 0.625 & -0.375 & -0.25 \\ -0.375 & 0.625 & -0.25 \\ -0.25 & -0.25 & 0.5 \end{bmatrix}$$

$$[C^{(2)}] = \begin{bmatrix} 0.75 & -0.25 & -0.5 \\ -0.25 & 0.417 & -0.167 \\ -0.5 & -0.167 & 0.667 \end{bmatrix}.$$

The next stage is to assemble the two separate elements together to form a global coefficient matrix. It can be seen in Fig. 7.17 that node 1 of element 1 and node 3 of element 2 do not share, while node 2 of element 1 and node 1 of element 2 contribute to the global node 2. Similarly, node 3 of element 1 and node 3 of element 2 constitute the global node 4. The situation is similar to the resistance matrix which describes the circuit in Fig. 7.9(a).

The global stiffness matrix is then:

$$[C] = \begin{bmatrix} C_{11}^{(1)} & C_{12}^{(1)} & 0 & C_{13}^{(1)} \\ C_{21}^{(1)} & C_{22}^{(1)} + C_{11}^{(2)} & C_{12}^{(2)} & C_{23}^{(1)} + C_{13}^{(2)} \\ 0 & C_{21}^{(2)} & C_{22}^{(2)} & C_{23}^{(2)} \\ C_{31}^{(1)} & C_{32}^{(1)} + C_{31}^{(2)} & C_{32}^{(2)} & C^{(1)}{}_{33} + C^{(2)}{}_{33} \end{bmatrix}$$

or

$$[C] = \begin{bmatrix} 0.625 & -0.375 & 0 & -0.25 \\ -0.375 & 1.125 & -0.25 & -0.75 \\ 0 & -0.25 & 0.417 & -0.167 \\ 0.5 & -0.75 & -0.167 & 1.167 \end{bmatrix}.$$

The expression for minimum work for the total system is obtained by differentiation for each variable in the system. This provides a set of simultaneous equations which can be solved for the unknowns. In our case this gives us

$$V_2 = -\frac{1}{C_{22}}(V_1 C_{12} + V_3 C_{32} + V_4 C_{42})$$

$$V_4 = -\frac{1}{C_{44}}(V_1 C_{14} + V_2 C_{24} + V_3 C_{34}).$$

The boundary conditions, $V_1 = 0$, $V_3 = 100$ are inserted into these equations which can now be solved iteratively to give $V_2 = 55.62$ and $V_4 = 50.07$.

Residual methods

The energy methods which have been presented above are derived from variational calculus and are often referred to as Rayleigh–Ritz techniques. However it is not always possible to derive a suitable functional to describe W^e. Since we are already using numerical methods, there seems to be no reason why we should not make greater use of approximations. In fact one of the most commonly used FEM approaches does this by the method of residuals. In the first instance an approximation or trial solution is devised. The difference between this and the true result is called a residual. We then seek to minimize the residuals over the entire system in the knowledge that when we have achieved this then we have the best approximation to the solution.

We will start with details of the mechanism of residual calculus. We will approach this by introducing the idea of trial functions and show by means of a staged example how trial functions and residuals can be used to solve differential equations. We will then conclude the chapter with a systems overview of how these can be applied in two-dimensional FEM.

(a) Trial functions

Let us suppose that we have a differential equation subject to boundary conditions which has a solution $V(x)$, then an approximation to the solution can be represented by the polynomial

$$\tilde{V}(x) = \phi_0(x) + a_1\phi_1(x) + a_2\phi_2(x) + \ldots$$

where $\phi_i(x)$ are trial functions and a_i are undetermined parameters (note: the tilde indicates that this is an approximate solution). The objective is to find expressions for $\phi_i(x)$ and this will be done as an example in two parts.

Example 7.6(a) Trial functions for a differential equation
We will solve the differential equation

$$x\frac{d^2V}{dx^2} + \frac{dV}{dx} = x^2$$

with boundary conditions $V(1) = 1$ and $\dfrac{dV(2)}{dx} = 2$ in the range $1 \le x \le 2$.

The appearance of the double derivative on the left side and the squared term on the right of this equation suggest that the trial function will take the form

$$\tilde{V}(x) = \alpha_1 + \alpha_2x + \alpha_3x^2 + \alpha_4x^3$$

where the α terms are yet to be determined.
The boundary conditions can be written as

$$\tilde{V}(1) = \alpha_1 + \alpha_2 + \alpha_3 + \alpha_4 = 1$$

and

$$\frac{d\tilde{V}(2)}{dx} = \alpha_2 + 4\alpha_3 + 12\alpha_4 = 2.$$

These two equations are insufficient for a complete solution, but allow us to eliminate two α terms from the boundary condition equations.

Choosing to eliminate α_1 and α_2, and using Maple to complete the algebra (collect(",[a$_3$, a$_4$], distributed); being particularly helpful), we obtain

$$\tilde{V}(x) = (1 + 2(x - 1)) + \alpha_3((x - 3)(x - 1)) + \alpha_4((x^2 + x - 11)(x - 1))$$

which is identical in form to eqn (7.8) i.e.

$$\tilde{V}(x) = \phi_0(x) + a_1\phi_1(x) + a_2\phi_2(x) + \dots .$$

A comparison of coefficients now yields the trial functions:

$$\phi_0(x) = 1 + 2(x - 1)$$
$$\phi_1(x) = (x - 3)(x - 1)$$
$$\phi_2(x) = (x^2 + x - 11)(x - 1)$$

so that the trial function is

$$\tilde{V}(x) = [1 + 2(x - 1)] + a_1[(x - 3)(x - 1)] + a_2[(x^2 + x - 11)(x - 1)]$$

(b) Residuals

The values for the trial functions in the example have been established, but the undetermined parameters of eqn (7.8) have not. The difference between the trial and true solution represents an error. By placing the trial solution into the differential equation and obtaining an expression for the difference we can obtain a measure of the error, which we call a *residual $R(x, a)$*.

Our objective is then to minimize this residual over the range of interest. There are many ways of doing this. One method is similar to the least squares method in statistics in that it attempts to minimize the mean square of the residual with respect to each parameter over the domain of interest. Other approaches include the Rayleigh–Ritz and the collocation methods. However the one which will be used here is the Galerkin or weighted residual method. We have chosen to present it here because it is perhaps the most widely used.

The Galerkin method treats the trial functions $\phi_i(x)$ as weights. If we have N values of the undetermined parameter (a_1 *to* a_N) then we have N weighted residuals of the form $R(x, a)\phi_i(x)(i = 1, \dots, N)$. If the region of interest is from $x = a$ to $x = b$ we then have a set of equations of the form:

$$\int_a^b R(x, a)\, \phi_i(x)\, dx = 0 \qquad (i = 1, \dots, N) \tag{7.9}$$

and these can be used to solve the undetermined parameters.

Example 7.6 (b) Residuals for Example 7.6 (a).
If we were to insert the trial solution (eqn (7.8)) into the differential equation of Example 7.6(a) we would obtain a residual:

$$R(x, a) = x\frac{d^2\tilde{V}}{dx^2} + \frac{d\tilde{V}}{dx} - x^2 \neq 0.$$

If we then insert the expression for the trial functions which were obtained in Example 7.6(a) we get:

$$R(x, a) = (2 - x^2) + 4\,(x - 1)\,a_1 + 3\,(3x^2 - 4)\,a_2.$$

This can then be used in eqn (7.9) over the specified range ($1 \leq x \leq 2$)

$$\text{Thus} \int_a^b R(x, a)\,\phi_1(x)\,dx = 0 \text{ is now}$$

$$\int_1^2 [(2 - x^2) + 4\,(x - 1)\,a_1 + 3\,(3x^2 - 4)\,a_2]\,[(x - 3)(x - 1)]\,dx = 0,$$

$$\text{and similarly} \int_a^b R(x, a)\,\phi_2(x)\,dx = 0 \text{ is}$$

$$\int_1^2 [(2 - x^2) + 4\,(x - 1)\,a_1 + 3\,(3x^2 - 4)\,a_2]\,[(x^2 + x - 11)(x - 1)]\,dx = 0$$

These can be solved to give $a_1 = -1.05$ and $a_2 = 0.27$.

A graph of the approximate solution which was obtained using this method is shown in Fig. 7.18. Within the limits of graphical error it is identical to the analytical solution.

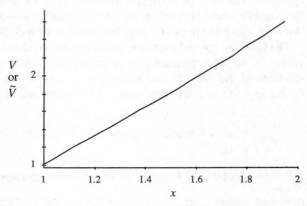

Fig. 7.18 Analytical and Galerkin weighted residual solutions for the differential equation. There is no discernible difference in the results over this range of x values.

Residual methods in two-dimensional FEM

The previous section has demonstrated how we can devise an approximate solution which is inserted into the governing differential equation in order to obtain a residual. When we apply these principles to FEM what we are doing is attempting to get the best fit to the exact solution of the system (a patchwork) of differential equations covering the region being modelled and subject to the relevant boundary constraints. Although the formulation is easily stated, the reality is considerably more complicated. Our presentation which is based on that of Burnett [7.4] will be largely a systems overview of the procedure. Burnett specifies a twelve-step procedure for FE analysis using residual methods:

1. write a Galerkin equation for a typical element;

2. integrate by parts;

3. put trial solution into integrals in residual equations;

4. develop expressions for the shape functions;

5. substitute shape functions into element equations;

6. prepare expressions for flux;

7. specify numerical data for the problem;

8. evaluate interior terms for each element and assemble into a system;

9. apply boundary conditions;

10. solve the system equations;

11. evaluate the flux;

12. display results and estimate the accuracy.

In this section we provide a brief discussion on the first four of these which are largely concerned with theoretial aspects which are of general relevance and follow on from previous sections. We will dwell a little on step 4 as it clearly confirms why FEM tends to use non-Cartesian discretizations. Anyone requiring further details is recommended to consult the cited reference.

1. A general expression for a two-dimensional boundary-value problem is given as:

$$-\frac{\partial}{\partial x}\left[\alpha_x(x,y)\frac{\partial U(x,y)}{\partial x}\right] - \frac{\partial}{\partial y}\left[\alpha_y(x,y)\frac{\partial U(x,y)}{\partial y}\right]$$
$$+ \beta(x,y)\,U(x,y) = f(x,y). \tag{7.10}$$

For example the equation for two-dimensional thermal convection

$$\frac{\partial q_x}{\partial x} + \frac{\partial q_y}{\partial y} + \frac{h}{t}(T - T_\infty) = Q \quad (q_x, q_y = \text{directional heat flux})$$

can be rewritten as

$$-\frac{\partial}{\partial x}\left[k_x \frac{\partial T(x, y)}{\partial x}\right] - \frac{\partial}{\partial y}\left[k_y \frac{\partial T(x, y)}{\partial y}\right] + \frac{h}{t}T(x, y) = Q + \frac{h}{t}T_\infty.$$

For this problem, a general trial solution of the form of eqn (7.8) can be given as:

$$\tilde{T}(x, y) = \phi_0(x, y) + a_1\phi_1(x, y) + a_2\phi_2(x, y) + \cdots$$

where $\phi_i(x, y)$ are trial functions and a_i are undetermined parameters.

If a typical element within a FE analysis of this problem is chosen and if the trial solution is inserted into the differential equation then we will obtain a residual $R(x, y; a)$

$$R(x, y; a) = -\frac{\partial}{\partial x}\left[k_x \frac{\partial \tilde{T}(x, y)}{\partial x}\right] - \frac{\partial}{\partial y}\left[k_y \frac{\partial \tilde{T}(x, y)}{\partial y}\right] + \frac{h}{t}\left[\tilde{T}(x, y) - T_\infty\right] - Q.$$

Then for this individual element we have a set of Galerkin equations, one for each degree of freedom (one for each undetermined parameter, a_i)

$$\iint_{\text{element}} R(x, y; a)\, \phi_i(x, y)\, dx\, dy = 0 \quad (i = 1, \ldots, N). \tag{7.11}$$

2. The expression for $R(x, y; a)$ appropriate to a particular problem (such as thermal convection) are substituted into eqn (7.11) and each component in the system of equations is integrated (not always an easy task).

3. In this step we take the set of integral equations which result from Step 2 and for each one the appropriate expression for the general solution ($\phi_0(x, y) + a_1\phi_1(x, y) + a_2\phi_2(x, y) + \ldots$) is inserted and the algebra is then simplified as far as possible. The end product at this stage is a set of equations of the form

$$K_{i1}a_1 + K_{i2}a_2 + \ldots + K_{iN}a_N = F_i \quad (i = 1, \ldots, N) \tag{7.12}$$

so that for the individual element we now have a matrix equation with the stiffness and load terms which are the concerns of finite-element modellers.

4. If the interior of a two-dimensional element is well behaved in a material sense then the shape functions will be amenable to a linear approximation. The approach is then identical to that which was outlined by eqns (7.4)–(7.7). If

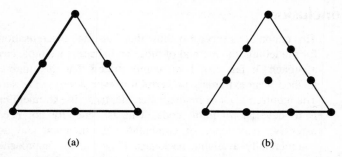

Fig. 7.19 (a) Quadratic and (b) cubic elements involving complete polynomials.

this is not the case then higher-order approximations may be necessary. We could always have a finer discretization so that linear approximations could be used within smaller elements, but this could involve a significant computational load. There are other approximation methods which seek to provide an accurate description of the response. One of these is the use of a polynomial expression. Such expressions can have low order (only a few terms) or can have many terms to improve the accuracy. Experience with FEM tends to show that it is better to have fewer (larger) elements and to use high-order polynomials. However one of the benefits of low-order approximations is that they always guarantee continuity across the interfaces so long as continuity is imposed at the apexes of the triangular elements.

The higher-order polynomials can be cubic, quartic, quintic, etc. and the outline shape of the first two of these for triangular elements are shown in Fig. 7.19. The appropriate expressions can be seen to have much in common with a Pascal triangle.

$$
\tilde{V}^e = \begin{array}{c}
a \\
+ bx + cy \\
+ dx^2 + exy + fy^2
\end{array} \quad \text{(quadratic)}
$$

$$
\tilde{V}^e = \begin{array}{c}
a \\
+ bx + cy \\
+ dx^2 + exy + fy^2 \\
+ gx^3 + hx^2y + hxy^2 + ky^3
\end{array} \quad \text{(cubic)} \tag{7.13}
$$

In each case these are complete polynomials in that they are the full description of the appropriate order. In situations where the polynomial is not complete then quadrilateral elements are preferred.

7.3 Conclusion

This chapter has served to show that Cartesian discretisations are not essential for modelling; the method of tubes and slices is a simple case of an alternative approach. It has also been demonstrated that the finite-element modelling method is an extremely powerful tool with a very wide range of applications. The required level of mathematical dexterity is extremely high if the modeller is to develop *ab initio* code. This is generally not necessary as there is extensive experience of computational efficiency and accuracy built into commercially available packages. However the approaches to FEM which have been presented in this chapter serve to explain some of the background and in particular why the minimization in numerical handling favours the use of triangular discretizations.

References

7.1   P. Hammond, *Energy methods in electromagnetism*, Oxford University Press 1981.
7.2   P. Hammond and J. Sykulski, *Engineering electromagnetism*, Oxford University Press 1994.
7.3   P.R. Silvester and R.L. Ferrari, *Finite elements for electrical engineers*, Cambridge University Press 1991.
7.4   D.S. Burnett, *Finite element analysis*, Addison-Wesley 1987.
7.5   A.M. Howatson, *Introduction to electrical circuits and systems*, Oxford University Press 1996.

Examples, exercises, and projects

7.1   The diagram in Fig. 7.20 shows a cross section of an electrical heater for a water distillation unit. Four elements are passing heat through an insulating sleeve $(k_T = 0.01 \text{ Js}^{-1}\text{cm}^{-1}\text{K}^{-1})$ to the surrounding boiling water. If the system were to be analysed using the tubes and slices technique devise a suitable meshing arrangement.

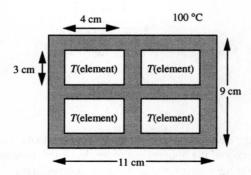

Fig. 7.20 Cross section of a heater.

If this system transmits heat to the water at 4 kW per metre of its length, calculate the equilibrium temperature of the elements.

7.2 The angle of turn on a rod under torsion can be described by the equation $q = TL/JG$ where T is axial torsion, J is polar moment of inertia, G is the shear modulus, and L is the length of the rod. A section of rod of length L can be treated as an element as shown in Fig. 7.21 and the behaviour can be expressed in matrix form as

$$\frac{J}{GL}\begin{bmatrix} 1 & -1 \\ -1 & 1 \end{bmatrix}\begin{bmatrix} \theta_1 \\ \theta_2 \end{bmatrix} = \begin{bmatrix} T_1 \\ T_2 \end{bmatrix}.$$

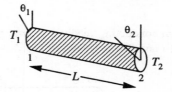

Fig. 7.21 A rod under torsion.

Consider a uniform rod of length $3L$ comprising three elements with global number 1, 2, 3, 4. If node 1 and 4 are fixed rigid and a torque T_a is applied at node 2 and a torque $-T_b$ is applied to node 3. Develop a global stiffness matrix and devise an expression for θ_2 and θ_3 as a function of T_a and T_b.

7.3 The shape in Fig. 7.15(a) is shown on the left in Fig. 7.22 with additional points along the periphery. Develop a suitable Voronoi diagram and hence a Delaunay triangulation.

The drawing on the right is the result of an arbitrary joining of the points. Use the circumcircle test to draw the best diagonals for each quadrilateral and compare the results with the Delaunay triangulation.

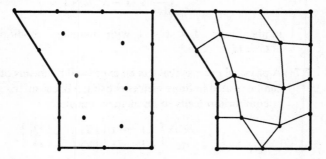

Fig. 7.22 A shape for triangulation.

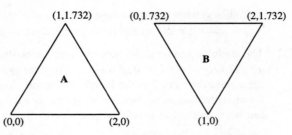

Fig. 7.23 Two triangular elements and their coordinates.

7.4 Use the coordinate data to compare the element shape functions and element coefficient matrices for the two trianglular elements shown in Fig. 7.23.

7.5 A uniform conducting medium is arranged to replicate the situation in Fig. 7.9(b) and this is shown in Fig. 7.24.

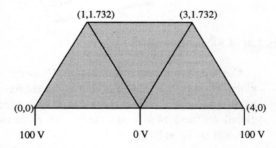

Fig. 7.24 Shape in a uniform conducting medium.

Use an approach similar to that in Example 7.5 to deduce the potentials at the centroids of each of the three elements.

7.6 Use the Galerkin weighted residual method to obtain a solution for the equation

$$\frac{d^2\theta}{dx^2} + (1 + x^2)\theta + 1 = 0$$

in the range $-1 < x < 1$ with boundary conditions $\theta(-1) = 0$ and $\theta(1) = 0$.

7.7 A physical process that can be expressed by means of a Laplace equation can be treated in finite elements using a linear matrix equation. Heat flow at equilibrium leads to an element equation

$$\frac{k_T A}{L} \begin{bmatrix} 1 & -1 \\ -1 & 1 \end{bmatrix} \begin{bmatrix} T_1 \\ T_2 \end{bmatrix} = \begin{bmatrix} Q_1 \\ Q_2 \end{bmatrix}.$$

That is: $\qquad$ const $[Y]\,[T] = [Q]$.

The matrix $[Y]$ expresses the conduction contribution of the element.

The matrix contribution of an element to thermal capacitance can be expressed as

$$\frac{C_p\,\rho L}{6}\begin{bmatrix} 2 & 1 \\ 1 & 2 \end{bmatrix}.$$

So that a parabolic equation for a single element, $\rho C_p \frac{\partial T}{\partial t} + k_T \frac{\partial^2 T}{\partial t^2} = 0$ can be expressed as

$$\frac{C_p\,\rho L.}{6}\begin{bmatrix} 2 & 1 \\ 1 & 2 \end{bmatrix}[\dot{T}] + \frac{k_T\,A}{L}\begin{bmatrix} 1 & -1 \\ -1 & 1 \end{bmatrix}[T] = [P]$$

which can be numerically integrated with respect to time to obtain a transient FEM analysis.

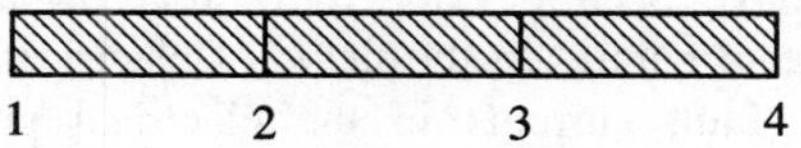

Fig. 7.25 A discretized copper rod.

The diagram in Fig. 7.25 represents a one-dimensional copper rod which is discretized into 3 elements each of 1 cm length. Initially all nodes are at 20 °C. At $t = 0$ the temperature of node 1 is raised and held at 100 °C while node 4 is held at 0 °C. Use the above information to develop a global matrix expression and use numerical integration to solve for $T(x = 2, t)$ and $T(x = 3, t)$. A time interval of 0.01 s might be appropriate.

Frequency domain models

Up to now in this book we have emphasized models in the time domain and have largely ignored the frequency domain except to say that it is possible to move from one to the other. In Chapter 4, we outlined the way that TLM modellers transfer to the frequency domain information by performing a discrete Fourier transform on the model output data. However we were not too specific about the nature of these techniques. The apparent imbalance will be redressed to some extent in this chapter.

The frequency domain is extremely important in its own right and many models have been developed to treat problems which are oscillatory or impulsive in nature. In the fields of electrical and electronic engineering for example, fault currents in the electrical power distribution network are oscillatory. Loudspeakers are a generic example of an oscillatory system which has a specific frequency response; tweeters will not perform the task of a woofer in a hi-fi system and vice versa. Such components have a characteristic frequency range and will even display resonances. It is possible to devise a circuit analogue which will (depending on its level of sophistication) account for some or all of the observed behaviour. A circuit simulator package such as SPICE can then be used to calculate the frequency response of the analogue circuit and this should accurately predict the behaviour of the loudspeakers. In a similar manner we can devise circuit analogues to simulate the performance of musical instruments. This was the basis of operation of electronic organs in the pre-digital era.

The material in Section 8.1 will be familiar to many electrical engineers and is therefore kept brief. Readers who require a more detailed introduction or who might wish to consolidate existing knowledge should consult the companion text book by Lancaster [8.1]. Once we have covered the fundamentals of frequency domain analysis, we will then concentrate on some of the better-known frequency domain modelling techniques and will support the presentation with several relevant examples.

8.1 Electrical and mechanical filters

Most readers may be aware of the hums and noises that are experienced within a moving motor vehicle. Motor manufacturers devote much effort to locating these and (depending on the price of the car) reducing them by means of damping, muffling, or possibly active noise control. The sources include impulsive sounds (such as from the exhaust). The exhaust pipe with silencer box is not unlike an organ pipe and will have standing waves along its length. Car manufacturers generally have rubber attachment points between the chassis and pipe so as to inhibit these. However, if the pipe should become detatched at one or more points, then it may start to vibrate (like a large tuning fork) and a modest family saloon can sound more like a high-performance car. The transmission system (engine, gear-box and prop shaft) cause vibrations and rumbles and some of these may introduce resonances within the passenger compartment. The nature of the adhesion between a car and the road surface can also cause frequency-domain effects: the road rumble from the tyres on a military vehicle is unmistakable. Finally there are fast-moving components within a car such as the alternator which can be the source of high pitched noises.

Analytical techniques and network analogues

The book by Fowkes and Mahoney [8.2] treats a range of examples concerned with vibrations in vehicles. They use analytical techniques which are mostly based on the second order differential equation

$$\frac{d^2y}{dt^2} = -\omega^2 y + A f(t). \tag{8.1}$$

We have already encountered several examples of the same equation earlier in this book. Equations of this form can be used to describe the behaviour of electrical circuits involving capacitors, inductors and resistors which can in turn represent the behaviour of mechanical systems. A spring, for example, has a natural frequency of oscillation and can be thought of as identical to a circuit comprising an inductor and capacitor in series. A dashpot or shock absorber can similarly be treated as a resistor. In our development of the concept of analogues, we will need to recall that we have considered the time dependent behaviour of capacitors and inductors (Chapter 4). Ohm's law is normally used to relate current and voltage through a resistor. However it can also be applied to inductors and capacitors where instead of resistance we talk of impedance. In these cases the results are imaginary numbers and are dependent on frequency. The capacitive and inductive impedances due to C and L are:

$$Z_C = \frac{1}{j\omega C}; \qquad Z_L = j\omega L; \tag{8.2}$$

where $j = \sqrt{(-1)}$ and ω is the angular frequency ($\omega = 2\pi f$, where f is frequency).

The equation for Z_C can be used in many ways. Real capacitors heat up when they are used in AC circuits. In theory, a perfect capacitor should not do

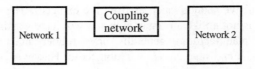

Fig. 8.1 A network representation of the coupled resonant circuits of Fig. 2.10.

this. This problem was considered in Example 2.5 where it was shown that Z_C was a complex quantity in non-ideal capacitors; it consisted of a real and an imaginary part. The real component contributes to heating. An almost identical analysis involving Z_L could be used to account for the fact that electrical transformers are non-ideal and heat up when used in AC circuits. This explains why neighbourhood electrical tranformers have extensive cooling fins or tubes. These impedances can also be applied to mechanical systems. The impedance Z_C might account for energy storage in a spring and Z_L can similarly be used to account for mechanical inertia.

Once we are happy that electrical entities such as R, L, C, V and I can be used as analogues, then we can talk about entire circuit analogues. In general terms we refer to these as networks. In Example 2.4 we showed how a circuit consisting of two coupled electrical resonators could explain the curious effect which pendulum clocks can have on each other, when placed close together on a wall. The observation can be then be described in general terms as the effect of coupling two networks (see Fig. 8.1).

Lancaster [8.1] gives details of the analysis of such connected networks by simply treating subsystems (e.g. the two clocks) as individual networks which can then be joined using the matrix methods which he describes.

A set of connected networks (in the frequency domain) can alternatively be considered as a cascade of electrical filters. This extension allows the techniques of filter theory to be applied to modelling.

We will now apply these concepts of networks and filter theory to some relevant examples.

Example 8.1 Filter models for an acoustic transducer
Acoustic transducers have played a critical role during two world wars, particularly in the area of submarine detection and location. *Passive* devices were the first to be developed. These consisted of piezoelectric crystals, which converted underwater pressure waves into electrical signals. They had terminals attached to opposite faces and registered a voltage which was proportional to the sound pressure at the face which was immersed in the water. They were used to 'hear' the presence of submarines. *Active* devices were developed at a later stage. If a large voltage impulse is applied to the terminals at opposite faces of a submerged transducer, then the piezocrystal

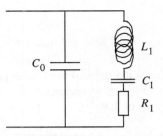

Fig. 8.2 van Dyke model of a transducer.

distorts and creates a disturbance which is transmitted through the water as a sound impulse. The time taken for the signal to impinge on a target and return as well as the frequency spectrum of the received signal provides comprehensive information about the target. This is the basis of *sonar*. In the following treatment we will start with a simple network model and we will then proceed in stages of increasing sophistication to treat an active transducer as a cascade of filters.

(a) van Dyke model

The response of an acoustic transducer to applied voltage is a function of frequency which does not depend linearly on amplitude because of internal resonances in the crystal. However, if the resonances are known in advance, electronic circuitry can be devised to overcome nonlinearities. Alternatively the designer may be able to adjust the position of the resonances so as not to interfere with readings. Early observations of the characteristics of piezocrystals led to a simple circuit for passive devices. This is called the van Dyke model and this is shown in Fig. 8.2.

The circuit of Fig. 8.2 can be replaced by a single impedance (Z_T) which represents the sum of R_1, L_1, and C_1 in series (designated as Z_{RLC}) which is in parallel with C_0. The parallel combination gives:

$$\frac{1}{Z_T} = \frac{1}{Z_{C_0}} + \frac{1}{Z_{RLC}}$$

or

$$Z_T = \frac{Z_{C_0} Z_{RLC}}{Z_{C_0} + Z_{RLC}}$$

where

$$Z_{C_0} = 1/j\omega C_0$$

and

$$Z_{\text{RLC}} = R_1 + j\omega L_1 + 1/j\omega C_1 = \frac{j\omega R_1 C_1 - \omega^2 L_1 C_1 + 1}{j\omega C_1}.$$

The final steps in the derivation of an expression for Z_T require care, and the use of a mathematical package with symbolic facilities (Mathematica, Maple, etc.) may be helpful.

(b) Characterization of two networks

It is well known that the transfer of energy between two networks is maximum when the impedances are matched; the 4 Ω output from an audio amplifier will give maximum output if a 4 Ω speaker is attached. Similarly, the impedance of a radio or TV antenna should be matched to the impedance of free space in order to receive the best signal. The same principles apply in the transfer of energy between an acoustic transducer and its surrounds, namely water. The transmission of sound through water can be modelled using electrical analogues as described in chapter 4. The distributed inductance (L_d) and capacitance (C_d) determine the value of Z_W (the impedance which the water presents to the transducer):

$$Z_W = \sqrt{\frac{L_d}{C_d}}.$$

It is generally the case that the water impedance and the transducer impedance are not the same, so that there will be a mismatch. If the observer is situated at the transducer looking into the water then the reflection coefficient will be:

$$\rho = \frac{Z_W - Z_T}{Z_W + Z_T}.$$

System characterization is important because at this stage we have a transducer, described by one network which is coupled to another network which models the water and it should be possible to simulate the transfer of signals. A received signal of high amplitude may be due to a significant external source or it may be due to resonances within the transducer.

(c) A multimode transducer model

The simple transducer which has been described above is not particularly efficient. The response of a crystal to an applied voltage impulse will cause a reaction on both faces of the crystal. Under normal circumstances only one face of the transducer is immersed in water while the other face is on the ship. Signals emitted from the in-board face served little or no purpose.

A significant design improvement can be achieved by attaching the rear face of the transducer to a section of metal such as Invar steel. Because of the nature of the elastic constants in this material there is very little mechanical distortion in response to an acoustic stress. The piezocrystal as a whole moves against an unyielding base and therefore a large proportion of the energy is transmitted into the water.

Improvements can also be achieved at the transmitting end. Since the material of the transducer is more dense than water, the value of ρ is very close to unity. This means that the water appears to be an open-circuit termination; most of the energy is reflected back from the transducer/water interface (it is precisely such a mismatch at the sea/air interface that explains why we do not normally hear underwater sounds). The level of matching can be increased by interposing one or more suitable materials (e.g. aluminium, plastic, or rubber) between the transducer and water. The geometry of such a yielding material can also be arranged so as to have a gradual impedance transition.

Although these developments have significantly increased the efficiency of acoustic transducers, the problem is now much more complicated than was outlined in the simple van Dyke model. The system comprising a sandwich of Invar tail-piece, piezocrystal, and head-piece has several mismatching interfaces. Sound reflected from one of these interfaces will travel back until it encounters another interface where there is yet another reflection. With lots of internal reflections there will be lots of resonances. In summary the problem looks like an ideal candidate for time-domain modelling using TLM (Chapter 4).

In the frequency domain the multiple internal reflections within the components of the sandwich can be viewed as a filtering process. This leads to an equivalent circuit model as shown in Fig. 8.3.

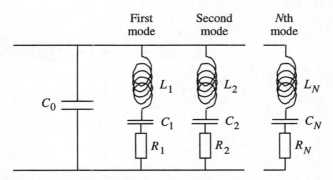

Fig. 8.3 A multimode lumped-parameter model for a transducer.

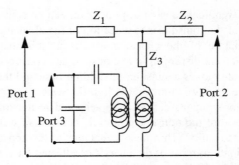

Fig. 8.4 Lumped-parameter model for a piezoelectric crystal. Ports 1 and 2 are connected to other sections of the transducer. Port 3 comprises the electrical terminals of the crystal and the transducer represents the conversion from electrical to mechanical signals (which we in turn will represent as electrical signals).

This situation can also be represented as the cascading of two-port networks. Figure 8.4 shows a lumped-parameter model for the section of a transducer sandwich which includes the piezocrystal.

Any further refinements of transducer modelling in the frequency domain requires a more complete transmission line description. The technique will be described in the next section before returning to the problem in Example 8.2.

8.2 Transmission lines in the frequency domain

Figure 8.5 shows the lumped-parameter model of a section of transmission line. If we consider a coaxial cable as our base example then R and L are the resistance and inductance of the central conductor, C is the capacitance which exists between the inner and outer conductors, and G is the leakage conductance of the intervening medium.

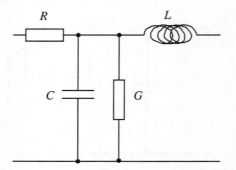

Fig. 8.5 A lumped-parameter representation of a section of transmission line.

The behaviour of the current and voltage on a transmission line is well known in electromagnetics and Lancaster [8.1] provides an excellent introduction to the subject. The governing equations are:

$$\frac{d^2V(z)}{dz^2} = \gamma^2 V(z) \qquad \frac{d^2I(z)}{dz^2} = \gamma^2 I(z) \tag{8.3}$$

where γ is called the *propagation constant* and is given by:

$$\gamma = \sqrt{(R + j\omega L)(G + j\omega C)} . \tag{8.4}$$

This is a complex parameter in that it contains real and imaginary parts. We can therefore write it as:

$$\gamma = \alpha + j\beta \tag{8.5}$$

where α is the attenuation and β is the phase lag.

Equation (8.4) is particularly unmanageable; it is not easy to relate $R, L, G,$ and C to α and β of eqn (8.5). We will therefore consider some special cases that are relevant to our models for acoustic transducers.

(a) Lossless transmission line $(R = G = 0)$

velocity, $v = \dfrac{1}{\sqrt{LC}}$

impedance, $Z_0 = \sqrt{L/C}.$

(b) Slightly lossy transmission line $(\omega L >> R, \omega C >> G)$

attenuation, $\alpha \sim \dfrac{1}{2}\left[\sqrt{\dfrac{C}{L}} + G\sqrt{\dfrac{L}{C}} \right]$

phase lag, $\beta \sim \omega\sqrt{LC}$

impedance, $Z_0 = \dfrac{L}{C}\left[1 + \dfrac{1}{2j\omega}(\dfrac{R}{L} - \dfrac{G}{C}) \right].$

(c) Leaky transmission line $(R = 0, G \neq 0)$

propagation constant, $\gamma = j\,\omega\,\sqrt{LC}\sqrt{1 + \dfrac{G}{j\omega C}} .$

It will be noted that almost all of the parameters in these specific cases are a function of frequency. Thus it should be possible to model the transmission line as a conventional circuit element and undertake a frequency response analysis. This approach is outlined for a modern design of underwater acoustic transducer in the next example.

Example 8.2 Transmission line model for a practical acoustic transducer
Figure 8.6 shows the schematic design of a practical transducer of
cylindrical geometry. It will be noted that the entire assembly is held
together by means of a prestressed bolt.

Each component in this device can be represented by means of a
transmission line section which attempts to model the acoustic properties in
that material. The velocity of sound in a section will be given by:

$$v = \sqrt{\frac{Y}{\rho}}$$

where Y is Young's modulus of the material and ρ is the density. The
piezoelectric section must include a transformer in order to allow
conversion between electrical and mechanical energy.

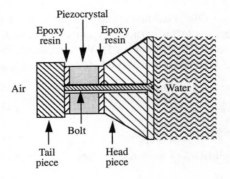

Fig. 8.6 A practical underwater acoustic transducer showing the prestressed bolt
which holds the assembly together.

Certain corrections should be applied in this particular case to deal with
frequency dispersion in the cylindrical sections. For the purposes of this
example we will consider only the most important one, namely the
Pochammer correction. This takes account of the fact that as a sound wave
travels in the z (axial) direction there are successive compactions and
rarefactions within the material. If a material has a significant *Poisson ratio*
then a compression in the axial direction will give rise to radial distortion.
The effective density of the material and hence the sound velocity is
altered.

Finally, the transmission-line sections are connected together in a way
which best describes the physical system. This is shown in Fig. 8.7. The
boxes marked 'air load' and 'water load' are the radiation impedances into
air and water respectively.

In order to use the model we translate the equivalent transmission-line circuit into an appropriate simulation package. We then apply a 'true' electrical input to the terminals shown in Fig. 8.7 and allow the transformer to convert these into electrical signals which form the input to the rest of the system. The package then calculates the frequency response of the overall system as well as a variety of other relevant parameters. Models of this type have now reached a high level of sophistication and are widely used by engineers to design efficient application-specific underwater acoustic transducers.

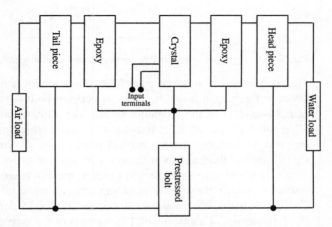

Fig. 8.7 Transmission-line circuit for the practical acoustic transducer shown in Fig. 8.6. Each box represents a line section with the propagation properties of that particular material over the frequency range under consideration.

The last example demonstrated how a successful model has been developed for a very important piece of equipment. It could also have been modelled using finite elements or some of the more modern variants of TLM code. However in each of these cases there would have been a need to include a more detailed description of the underlying physical processes. The frequency domain approach ignores much of the physics and provides engineers and designers with the type of information that they require. There are other circumstances where the frequency domain may be more appropriate simply because of the absence of physical data. The behaviour of the basilar membrane within the ear (our next example) is just such an example where *in vitro* measurements of physical parameters may not have any relevance to *in vivo* performance.

Example 8.3 A model for the basilar membrane (and semiconductor lasers)

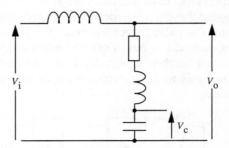

Fig. 8.8 A single transmission-line section of the basilar membrane model.

The cochlea within the ear is one of the most complex and least understood organ in the human body. It seems to perform both spectral analysis and neural encoding of the incoming sound. The strucure consists of a fluid-filled canal with small hairs (basilar membrane) which range in size from one end to another. The mechanical parts of the ear inject the acoustic signal into the fluid and as it travels it causes one or more small hairs to move. This has led researchers to develop models based on a cascade of resonators which convert the incoming pressure wave into basilar motion. The transmission line model outlined here is due to Ambikairajah *et al.* [8.3]. It consists of a cascade of 128 sections of the type shown in Fig. 8.8 where the following analogies apply

- series inductance ≡ fluid mass
- shunt inductance ≡ membrane mass
- resistance ≡ membrane losses
- capacitance ≡ membrane compliance.

Each section is designed for a different frequency and there are two important parameters which can be calculated using exactly the same approach as was used in the previous example: V_o/V_i represents the pressure transfer through a section; V_c is proportional to displacement of the membrane and is a direct measure of the strength of the neural signal transmitted to the brain.

Because of the very large number of sections it is computationally tedious to calculate the entire system and the authors of this model discuss methods of isolating each section so that it can be calculated independently of the others. The problem would seem to be ideal for a TLM treatment. Link transmission lines (see Section 4.3.2) at the interface between sections will ensure such isolation. Network analogues can then be devised to account for the frequency behaviour within each section and in this context it may be worth revisiting Section 4.2.4, (c) where we began modelling electrical filters and circuits. Observations can then be made in the time domain or can be transformed into the frequency domain using the techniques which will be discussed in Section 8.3.2.

Lowery [8.4] has used an identical approach to simulate the behaviour of semiconductor lasers. He ignores the complexities of semiconductor transport models which normally describes the physics at the level of electrons and holes. Instead, he divides the physical space into a series of isolated lossy resonators. These sections can also be interpreted as bidirectional two-port networks or filters. Optical or electrical pumping of the laser is then equivalent to the excitation of the cascade of networks. There will be a delay before there is an electrical output; analogous to the creation of a population inversion in the physical model. The initial output from the model may contain many frequency components before an equilibrium state is reached. Alterations can be made to the components within the individual sections in order to achieve the required output (e.g. narrow frequency spectrum, noise immunity, etc.). These have a direct physical interpretation and allow designers to quickly tailor the structure of a laser to a specific application.

8.3 Fourier techniques

Having introduced the frequency domain and shown how conventional circuit analysis can be used to model systems, we now move on to consider some other tools which are essential in frequency-domain analysis. In particular, we will be concerned with Fourier series and Fourier transforms. These have a long and honourable pedigree within the field of numerical methods. Indeed, there was a time when they were amongst the only available techniques for solving hyperbolic and parabolic differential equations. We will look at some examples from this perspective.

8.3.1 Fourier series

The principle of Fourier series (previously mentioned in eqn (2.9)) states that any function $h(t)$ that is periodic in the range 0 to 2π can be expressed as:

$$h(t) = A_0 + \sum_{n=1}^{\infty} (A_n \cos(nt) + B_n \sin(nt)) \tag{8.6}$$

where (assuming that the function is well behaved)

$$A_0 = \frac{1}{2\pi} \int_{-\pi}^{\pi} h(t) dt$$

$$A_n = \frac{1}{\pi} \int_{-\pi}^{\pi} h(t) \cos(nt) \, dt$$

$$B_n = \frac{1}{\pi} \int_{-\pi}^{\pi} h(t) \sin(nt) \, dt .$$

Thus a square wave described by $f(t) = -1(-\pi < t < 0)$, $f(t) = 1(0 < t < \pi)$ can be described by eqn (8.6) with the following values for the coefficients:

$$A_0 = 0; \qquad A_n = 0; \qquad B_n = \frac{2}{n\pi}(1 - \cos(n\pi)) = \frac{2}{n\pi}(1 - (-1)^n).$$

Now $B_n = 4/(n\pi)$ for n odd and $= 0$ for n even, Therefore

$$f(t) = \frac{4}{\pi}[\sin(t) + \frac{\sin(3t)}{3} + \frac{\sin(5t)}{5} + \ldots].$$

The level of the approximation to the analytical square wave then depends on the number of terms which are included in this series expression for $f(t)$.

Similarly, if we have a triangular wave defined by $f(t) = x(0 < x < \pi)$ and $f(t) = -x(-\pi < x < 0)$, then the Fourier series is:

$$f(t) = \frac{\pi}{2} - \frac{4}{\pi n^2} \cos(nt)(n = 1, 3, 5, \ldots).$$

We will now consider three examples where Fourier series can be used in order to obtain a solution.

Example 8.4 The response of a mass, spring, and damper mechanical system when subjected to periodic triangular excitation
The instantaneous behaviour of the arrangement shown in Fig. 8.9 can be described by the equation:

$$\frac{d^2z}{dt^2} + 0.02\frac{dz}{dt} + 25z = F(t)$$

where the excitation $F(t)$ is periodic triangular wave and is defined by:

$$F(t) = t + \pi/2 \text{ from } -\pi < t < 0$$

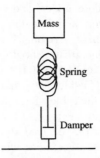

Fig. 8.9 A mass attached to a series arrangement of spring and damper.

$$F(t) = -t + \pi/2 \text{ from } 0 > t > \pi.$$

Now $F(t)$ can be converted into its equivalent Fourier series:

$$F(t) = \frac{4}{n^2\pi}\cos nt \ (n = 1, 3, 5, \ldots).$$

The differential equation can then be restated as

$$\frac{d^2z}{dt^2} + 0.2\frac{dz}{dt} + 25z = \frac{4}{n^2\pi}\cos(nt) \ (n = 1, 3, 5, \ldots).$$

The solution of the differential equation will have the form

$$z_n = A_n \cos(nt) + B_n \sin(nt).$$

This can be substituted into the previous expression to give:

$$A_n = \frac{4(25 - n^2)}{n^2\pi[(25 - n^2)^2 + (0.02n)^2]}; \quad B_n = \frac{0.08}{n\pi[(25 - n^2)^2 + (0.02n)^2]}.$$

Therefore in the time domain the solution to the differential equation, $z(t)$ will be the sum of terms of all z_n (where n is odd). Of course it is not reasonable to sum all terms up to $n = \infty$ and since the excitation falls away as $1/n^2$ we will make an approximation by summing up to a value of n that provides the required level of accuracy.

Example 8.5 Kelvin arrival curves and heat conduction
Earlier in this chapter it was mentioned that there was a time when Fourier methods were the only available method for solving certain classes of differential equations. This example clearly illustrates the point. In 1857 when it was proposed to lay a transatlantic telegraph cable, there was much controversy about the feasibility of such an enterprise. The nature of

electricity was little understood. Some believed that the 'electric fluid' would be squeezed out of existence at the ocean depths at which the cable was to be laid. There were others who believed that the greater the distance to be traversed, the greater should be the initial voltage to impel the signal on its way. William Thomson (later Lord Kelvin) took a different view. He argued that an RC ladder network of the form shown in Fig. 8.10 provided a good model of the cable (these days we might recognize this as a cascade of low-pass filters). He applied Fourier methods to analyse this circuit and his predictions of the received signal as a function of time (arrival curve) was in very close agreement with what was eventually observed. This contributed significantly to the understanding of electricity and provided cable manufacturers with a clear set of design criteria which have been in use ever since.

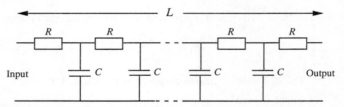

Fig. 8.10 One-dimensional RC ladder network.

We might also recognize the ladder network of Fig. 8.10 as a good model for heat-flow simulations as outlined in Chapter 4. In the next example we will show how Fourier methods lend themselves in a natural way to the analytical solution of problems in heat flow and matter diffusion.

Example 8.6 Heat flow and matter diffusion
We will start this example by restating the diffusion equation alongside its electrical analogue and our analysis will concentrate on the latter.

$$D\frac{\partial^2 \Phi(x,t)}{\partial x^2} = \frac{\partial \Phi(x,t)}{\mathrm{d}t}; \quad \frac{\partial^2 V(x,t)}{\partial x^2} = R_\mathrm{d}C_\mathrm{d}\frac{\partial V(x,t)}{\mathrm{d}t};$$

where (Φ is temperature for heat flow and concentration for matter diffusion).

The voltage in the electrical analogue can be separated into two variables:

$$V(x,t) = F(x)G(t).$$

If we have $V(0, t) = V(L, t) = 0$ as boundary conditions and we start with an initial profile within the sample described by

$$V(x, 0) = f(x)$$

then the separation of variables leads to two equations in term of a variable k. We might now recognize k from elsewhere as the wave number $(= 2\pi/l\lambda)$:
i.e. in order for

$$V(x, 0) = f(x) = \sum_{n-1}^{\infty} B_n \sin(\frac{n\pi}{L}x)$$

to satisfy the above expression for $V(x, t)$ we must define B_n as the half range expansion for the Fourier series of $f(x)$.
Thus

$$B_n = \frac{2}{L} \int_0^L f(x) \sin(\frac{n\pi}{L}x) \, dx$$

and

$$V(x, t) = \sum_{n=1}^{\infty} \left[\frac{2}{L} \int_0^L f(x) \sin(\frac{n\pi}{L}x) \, dx \right] \sin(\frac{n\pi}{L}x) \exp(-\lambda_n^2 t).$$

Although this last example seems to have strayed back into the time domain, it does demonstrate that processes which can be physically modelled using a cascade of low-pass filters can also be interpreted in terms of Fourier components, i.e. in the frequency domain.

8.3.2 Fourier transforms

The Fourier transform is a mathematical process which allows exchange of information between the time and frequency domains. Let us suppose that we have a period function of time $h(t)$. We can derive an equivalent function in frequency $H(f)$ by means of the following:

$$H(f) = \int_{-\infty}^{\infty} h(t) e^{2\pi j f t} dt. \tag{8.7}$$

For example, if we have an event such as an electromagnetic disturbance whose variation as a function of time was given by $h(t) = e^{-at^2}$, then the spectrum of frequencies which constitute this could be derived from eqn (8.7) as:

$$H(f) = \frac{1}{\sqrt{2a}} e^{-(2\pi f)^2/4a}.$$

If, on the other hand we have a frequency function $H(f)$, then it is possible to derive the time function by means of the inverse Fourier transform:

$$h(t) = \int_{-\infty}^{\infty} H(f)e^{-2\pi j f t} df. \tag{8.8}$$

So, if a frequency spectrum is defined by $H(f) = e^{-2\pi Tf}$ where $T > 0$, then

$$h(t) = \sqrt{\frac{2}{\pi}} \left[\frac{T}{t^2 + T^2} \right].$$

The discrete Fourier transform

In most of the applications in this book we deal with time discrete or sampled data; the output at the end of an iteration. Such data was highlighted in the chapter on TLM, where we mentioned that Johns and Beurle [8.5], who had pioneered the technique, transformed their time domain results to the frequency domain to confirm the results of their new method. The Fourier transform can also be applied to discrete time domain data derived using finite difference models (Chapter 3) and rule-based models (Chapter 5).

In order to formalize the discrete Fourier transform (DFT) we must first present some parameters for time discrete data sets. These include the sampling interval Δ and its reciprocal, the sampling frequency. In signal processing [8.6] the maximum frequency contained in an analogue signal is referred to as the Nyquist frequency f_c. It can be considered as a ceiling in our frequency domain because the theoretical minimum sampling rate which is necessary to allow the analogue signal to be recovered is $2f_c$. We can therefore define the *Nyquist frequency* as:

$$f_c = \frac{1}{2\Delta}. \tag{8.9}$$

In the present context this means that in order to get any meaningful information from our data we must restrict ourselves to frequencies of less than f_c.

If data is uniformly sampled then we can represent the time-domain values in an algebraic form as:

$$h_k = h(t_k) \quad \text{where } t_k = k\Delta \ (k = 0, 1, \ldots, N-1). \tag{8.10}$$

Note the use of the lower-case 'h' which distinguishes this as time-domain data as in eqn (8.7). The upper-case 'H' represents transformed (frequency-domain) data.

As it is presented eqn (8.7) covers the range of all frequencies. However, since the time-sampled data, $h(t_k)$ of eqn (8.10) is restricted (it does not contain frequency components above f_c) then the available values of $H(f)$ range from $-f_c$ to f_c. In other words, after transformation of time discrete data we obtain a restricted set of discrete frequency values:

$$f_n = \frac{n}{N\Delta} \text{ where } n = -\frac{N}{2}, \ldots, \frac{N}{2}.$$

The integral in eqn (8.7) is therefore aproximated by a sum

$$H(f_n) = \sum_{k=0}^{N-1} [h_k e^{2\pi j f_n t_k})]\, \Delta \tag{8.11}$$

$$= \Delta \sum_{k=0}^{N-1} h_k\, e^{2\pi j \frac{n}{N\Delta} k\Delta}$$

$$= \Delta \sum_{k=0}^{N-1} h_k\, e^{2\pi j k \frac{n}{N}}$$

$$= \Delta\, H_n.$$

This expresses the *discrete Fourier transform* (DFT)
There is similarly an expression for the *inverse DFT*

$$= \frac{1}{N} \sum_{k=0}^{N-1} H_n\, e^{-2\pi j k \frac{n}{N}}$$

$$= \Delta H_n.$$

At this stage it is quite useful to review the DFT using trigonometrical expressions. We know from Section (8.3.1) that any continuous periodic signal can be represented by a sum of sines of different amplitude and phase

$$h(t) = \sum_x D_x \sin(2\pi f_x t + \phi_x) \tag{8.13}$$

$$= \sum_x (E_x \sin(2\pi f_x t) + F_x \cos(2\pi f_x t))$$

where $E_x = D_x cos(\phi_x)$ and $F_x = D_x \sin(\phi_x)$. If the same periodic signal is sampled then eqn (8.13) can be written as:

$$h(n) = \sum_x (E_x \sin(2\pi \frac{f_x}{f_c} n) + F_x \cos(2\pi \frac{f_x}{f_c n})) \tag{8.14}$$

where $n = 0, 1, 2, \ldots, N$ and f_c is the Nyquist critical frequency.

For *any* summation of this form the following property is true:
If

$$Y = \frac{1}{N}\sum_{N=0}^{N-1}(E\,\sin(2\pi x n) + F\,\cos(2\pi x n))(n = 0, 1, 2\ldots)$$

then

$$Y = F \text{ if } x \text{ is an integer } and \text{ } N \text{ is large}$$

(8.15)

$$Y = 0 \text{ if } x \text{ is not an integer } and \text{ } N \text{ is large.}$$

This 'mathematical filtering' property can be used as follows: suppose we are interested in a frequency $f_?$; we would like to know the magnitude of its contribution to our time domain data. In order to obtain this information we multiply eqn (8.14) with an integral sinusoidal contribution of $f_?$.

$$\frac{1}{N}\sum_{N=0}^{N-1}\left[\sum_x (E_x\,\sin(2\pi\frac{f_x}{f_c}n) + F_x\,\cos(2\pi\frac{f_x}{f_c}n))\right]\sin(2\pi\frac{f_?}{f_c}n).$$

(8.16)

When this is multiplied out we get:

$$\frac{1}{N}\sum_{N=0}^{N-1}\sum_x\left[E_x\left[\cos(2\pi\frac{f_x-f_?}{f_c}n) - \cos(2\pi\frac{f_x+f_?}{f_c}n)\right]\right.$$
$$\left. + F_x[\sin(2\pi\frac{f_x+f_?}{f_c}n) - \sin(2\pi\frac{f_x+f_?}{f_c}n)]\right].$$

(8.17)

From eqn (8.15) we can see that this expression tends to zero as N becomes large unless $(f_x \pm f_?/f_c)$ is an integer. However, since both f_x and $f_?$ are less than or equal to f_c. this mean that f_x must equal $f_?$. In this case eqn (8.14) will become equal to $E_?/2$.

Thus the effect of multiplying the stream of time-domain data by the factor $\sin(2\pi\frac{f_?}{f_c}n)$ is to filter out everything *except* the contribution from $f_?$. The resultant is half the amplitude of the corresponding frequency component. The same process can be repeated for all frequencies within the range of interest.

It can be similarly shown that if the time domain data is multiplied by $\cos(2\pi\frac{f_?}{f_c}n)$ then the non-zero resultant is $F_?/2$, half the magnitude of the cosine term in eqn (8.14) for the particular frequency $f_?$.

The fast Fourier transform

The techniques described above are extremely useful but they are highly inefficient if used on very large sets of data. If there are N data points in time

Table 8.1 Number of multiplications required for DFT
and FFT

N	N^2	$N \log_e N$
10	10^2	46
10^2	10^4	92
10^3	10^6	138
10^4	10^8	184

and k points of interest in frequency space, then their elucidation appears to require a total of N^2 multiplications. The process can be improved if we can impose certain conditions on the data. This is because it is possible to take any Fourier transform of length N and present it as the sum of two transforms (one odd and one even) of length $N/2$. Now each of these can be similarly reduced to transforms of length $N/4$ and so long as N is some power of 2, the process can be continued recursvely until we have a sum of transforms each of length 1. This is the basis of the fast Fourier transform (FFT) and the number of multiplications which are required in order to get $H(f_n)$ from N time-domain data points is $N \log_e N$. The difference in the number of calculations can be seen in Table 8.1. It becomes quite clear that a large mesh problem which would be computationally enormous for the DFT is quite trivial for the FFT.

The actual details of the development of FFT algorithms are outside the scope of this book, but references [8.7] and [8.8] might be useful. Codes for DFT and FFT are also available in Mathematica and Maple.

Example 8.7 Speech analysis/synthesis and the Fourier transform
Over many years there has been considerable effort in developing and refining speech-synthesis equipment. There has also been much work on speech recognition; telecommunications companies see the benefits of automating such services for directory enquiries. Speech synthesizers have great benefits for those who have lost the power of speech; the case of the physicist Stephen Hawking is perhaps amongst the best known. Nevertheless both recognition and synthesis techniques have a long way to go before machines can interpret conventional speech and/or sound just like a person. They are very large and active areas of research and any presentation must of neccessity be selective. This example concentrates on the initial phases in the evolution of some speech models, namely the collection and interpretation of data. We will look first at recognition and then discuss one approach to synthesis.

Most speech research is conducted in the frequency domain and is concerned with isolated sounds. It is easy to connect a microphone/ amplifier to the input of an oscilloscope and obtain time domain data as

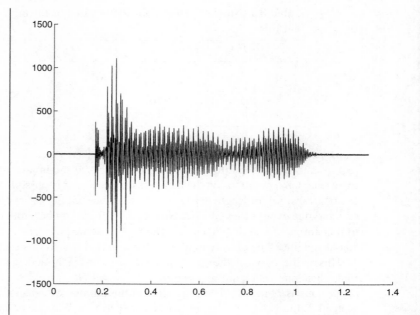

Fig. 8.11 Oscilloscope output of the utterance 'w' by a male speaker.

shown in Fig. 8.11. This is the equivalent of $h(t)$ in eqn (8.7). The advent of modern electronics has provided researchers with some very valuable tools. Fast sample-and-hold circuits combined with analogue-to-digital conversion means that the incoming sound is collected at a repetition rate Δ and is then stored as a data set, $h(t_k)$.

It is normal to collect these into time blocks (each with a power-of-2 number of points). Conversion to the frequency domain is accomplished using the FFT and this yields amplitude/frequency information equivalent to $H(f_n)$ of eqn (8.11) for each data set. The time blocks can then be reassembled to give what is called a 'spectrogram'. The frequency is plotted against time with the amplitude at any frequency presented either as colour or as grey level. The examples shown in Fig. 8.12(a),(b) are for the letter 'w' uttered by two different speakers.

It can be seen that the spectrograms for both speakers in Fig. 8.12 have certain characteristics in common. There is commonality in nearly all spoken utterances. Much current research is concerned with the rapid identification and interpretation of these features. In this respect the technique is analogous to optical character recognition (OCR) for the

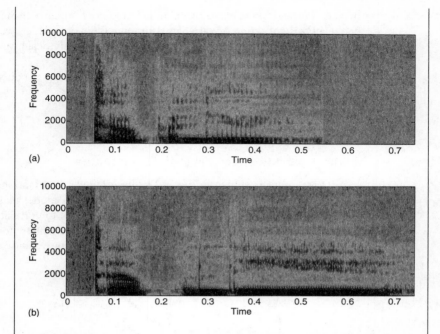

Fig. 8.12 Spectrograms of the letter 'w' uttered by (a) a male speaker and (b) a female speaker. The vertical axis is frequency and the horizontal axis is time. The amplitude is shown on a grey scale.

interpretation of computer scanned images. Speech intepreters are still at the early stages of development. Because machines continue to have problems with 'connected' sounds, most systems operate on a very restricted set of utterances. The telephone answering service 'BT Callminder' is a commercially operational example.

Several different approaches to speech synthesis are currently under development [8.9, 8.10]. Most rely heavily on aspects of speech and information theory which are really outside the scope of this book. Nevertheless we will include some details of a TLM model of the vocal tract due to Loasby [8.11] which is capable of producing isolated sounds.

Loasby's model treats the larynx as a generator in an acoustic network consisting of a transmission line (the throat) which splits into two lossy resonant cavities (the nasal region and the mouth). Although this is a simple system to describe in theory, the practical reality is much more complex. As we speak the sound generator is continuously changing in frequency and amplitude. At the same time the conformation of the throat, the size of the two cavities and the opening of the mouth and nose are also changing. At one level this collection of variables determine the words which we speak. At another, they determine the characteristics which give each of us our distinctive 'voice'.

In electrical analogue terms we assert that the following parameters are time dependent:

- The impedances of the transmission line sections

- The radiation characteristics from mouth and nose

- The resonant frequency and Q of the two cavities

A TLM circuit is shown in Fig. 8.13; the original Loasby arrangement had 4 transmission lines in each section. Although this model may never result in an effective speech synthesis system, attempts to correlate time-dependent physiological data with the model and attempts to match spectrograms for isolated words may lead to a better equivalent electrical circuit description for the vocal tract.

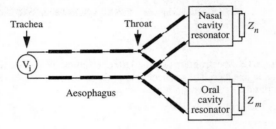

Fig. 8.13 A Loasby TLM model for the vocal tract. The values of impedance of each transmission-line section can change as a function of time. The value of Z_m, the mouth impedance, depends on the shape of the mouth. As in human physiology, there is also the possibility that the nasal impedance can have some variability.

Other speech models

The techniques which we have outlined in the previous sections are generally based on a model which uses physical analogues. There is an entirely different class of frequency domain models which use mathematical techniques. These include LPC (linear predictive coding of speech) [8.10] and HMM (hidden Markov model) [8.12].

LPC regards the vocal tract as a linear filter and uses automatic methods for estimating the transfer function and the excitation from 20 ms segments of a speech signal. It has enabled telecommunications companies to achieve large reductions in the bit rate required for speech transmission. The standard way of digitally coding speech signals was developed in the 1960s and is known as

Pulse Code Modulation (PCM). PCM requires 64 000 bits of information per second to be transmitted for telephone quality speech. Using LPC, a comparable quality can currently be achieved at a rate of 13 000 bits per second. The technique also has applications in speech recognition, because the filter transfer function and excitation are characteristic of the sound being uttered and can be used to recognize speech patterns.

HMM treats speech as a stochastic process [8.13]. Using a large quantity of labelled 'training' data, a set of HMMs of different speech sounds can be built and can then be used later to 'decode' an input speech signal into the most likely (in a probability sense) sequence of words spoken. Although this way of modelling speech sounds has been criticized as being too crude to encapsulate the subtleties of speech, its great advantage lies in the fact that it is capable of automatically processing very large amounts of speech data from thousands of different talkers. This enables the models to reflect the variations in speech signals caused by such factors as the speaker's size, accent, rate of speaking, etc. and so they can be used to recognize speech from a large population of different speakers, even when speaking over telephone lines. Over the past ten years, the technique has been considerably refined to the point where laboratory systems now exist which are capable of recognizing fluently spoken speech from many different speakers and with a vocabulary of several thousand words [8.14].

Acknowledgement

The assistance of S.J. Cox in the preparation of the section on speech processing is gratefully acknowledged.

References

8.1   G. Lancaster, *Introduction to fields and circuits*, OUP 1992.
8.2   N.D. Fowkes and J.J. Mahoney, *An introduction to mathematical modelling,* John Wiley and Son, Chichester 1994.
8.3   E. Ambikairajah, N.D. Black and R.L. Linggard, Digital filter simulation of the basilar membrane. *Computer Speech and language* **3** (1989) 105–118.
8.4   A.J. Lowery, Transmission line modelling of semiconductor lasers: the transmission line laser model. *International Journal of Numerical Modelling,* **2** (1989) 249–265.
8.5   P.B. Johns and R.L. Beurle, Numerical solution of 2-dimensional scattering problems using a transmission line matrix. *Proceedings of the IEE* **118** (1971) 1203–1208.
8.6   P.A. Lynn and W. Fuerst, *Introductory digital signal processing with computer applications* (2nd edition), John Wiley and Sons, Chichester 1994.
8.7   W.H. Press, S.A. Teukolsky, W.T. Vetterling, and B.P. Flannery, *Numerical Recipes in C*, Cambridge University Press 1992 pp. 496–608.
8.8   H.B. Wilson and L.H. Turcotte, *Advanced mathematics and mechanics applications using Matlab*, CRC Press 1994 pp. 177–213.

8.9   R. Linggard, *Electronic synthesis of speech* Cambridge University Press 1985.

8.10 B.S. Atal and S.L. Hanauer, *Speech analysis and synthesis by linear prediction of the speech wave. Journal of the Acoustical Society of America* **50** (1971) 637–655.

8.11 T.E. Cross, P.B. Johns and J.M. Loasby, *Transmission line modelling of the vocal tract and its application to the problem of speech synthesis.* Proc. IEE Conf. Speech Input/Output: Techniques and Applications No. 258, March 1986, 71–76.

8.12 S.J. Cox, Hidden Markov models for automatic speech recognition: theory and application. In *Speech and language processing*, C.Wheddon and R.Linggard (eds) Chapman and Hall, 1990.

8.13 D.R. Cox and H.D. Miller, *The theory of stochastic processes,* Chapman and Hall, 1965.

8.14 P.C. Woodland, C.J. Leggetter, J.J. Odell, V. Valtchev and S.J. Young, *The 1994 HTK large vocabulary speech recognition system,* Proceedings of the 1995 IEEE Conference on Acoustics, Speech and Signal Processing (ICASSP), Detroit.

Examples, exercises, and projects

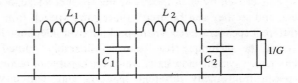

Fig. 8.14 Circuit for an LC ladder network.

8.1 The frequency response of a filter such as an LC ladder network can be treated as follows (see Fig. 8.14).

The relationship between the input current/voltage and the output current/voltage can be derived from a matrix analysis

$$\begin{bmatrix} V_i \\ I_i \end{bmatrix} = \begin{bmatrix} 1 & sL_1 \\ 0 & 1 \end{bmatrix} \begin{bmatrix} 1 & 0 \\ sC_1 & 1 \end{bmatrix} \begin{bmatrix} 1 & sL_2 \\ 0 & 1 \end{bmatrix} \begin{bmatrix} 1 & 0 \\ sC_2 + G & 1 \end{bmatrix} \begin{bmatrix} V_o \\ I_o \end{bmatrix}$$

where $s = j\omega$ ($j = \sqrt{(-1)}$ and $\omega = 2\pi f$). This cascade of matrices can be multiplied out to give a matrix of the form

$$\begin{bmatrix} V_i \\ I_i \end{bmatrix} = \begin{bmatrix} A_{11} & A_{12} \\ A_{21} & A_{22} \end{bmatrix} \begin{bmatrix} V_o \\ I_o \end{bmatrix}.$$

If I_o is set equal to zero then the relationship between the input and output voltages is given by:

$$\frac{V_o}{V_i} = \frac{1}{A_{11}}.$$

This treatment can be extended for a ladder of as many sections as desired.

In the first instance this analysis should be applied to obtain the filtering effect of a ladder LC network when presented with a 'white noise' input which is clipped to a maximum of ± 1 V. The component values are $C = 10 \mu f$ $L = 1$ mH, $R = 1000 \Omega$.

The analysis can then be extended to use a Kennedy-type Monte Carlo approach (see Section 6.3.1) to consider the frequency response of large batches of such filters which are manufactured using standard components which have $\pm 20\%$ tolerances. Such components are manufactured so that they have a Gaussian distribution about the nominal value. They are then tested and those that are outside the 20% limit are discarded. Thus the constituent components of the filter will exhibit truncated Gaussian distributions.

As a first stage of this analysis it might be worth starting with fixed value components and allowing the value of only one to vary. It should be observed that the overall effect of variations in a particular inductor or capacitor depend on its position in the ladder. Might this have a bearing on the design specification in such filters?

8.2 Several years ago there was a piece of publicly available software called *Geoclock* which plotted the position of the sun on a projection of the earth. At equinox, when the sun is directly over the equator, the insolation profile looks like a square wave. During winter in the northern hemisphere the profile looks more like that shown in the diagram of Fig. 8.15.

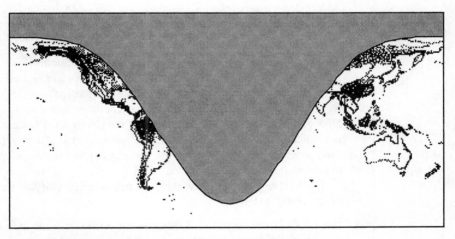

Fig. 8.15 Winter insolation profile for the northern hemisphere.

Now imagine these profiles to be the $0 - 2\pi$ definitions of 'continuous' waveforms. The Fourier components of a square wave (equinox) are well known. You should try to determine the components corresponding to the profiles at other times of the year and hence plot the variation of individual components as a function of time of year.

8.3 Example 8.6 covered the derivation of a solution of a heat flow or matter diffusion problem in one dimension using a Fourier approach. For planar diffusion this can be easily extended to two dimensions. Consider the problems of variable separation in spherical and cylindrical problems.

The diffusion equation in a sphere can be written as:

$$\frac{\partial^2 N(r,t)}{\partial r^2} + \frac{2}{r}\frac{\partial N(r,t)}{\partial r} = \frac{1}{D}\frac{\partial N(r,t)}{\partial t}.$$

It can be shown that the transformation $U(r,t) = N(r,t)r$ is particularly useful and leads to an expression:

$$\frac{\partial^2 U(r,t)}{\partial r^2} = \frac{1}{D}\frac{\partial U(r,t)}{\partial t}.$$

The derivation of a solution for $U(r,t)$ is assisted by the use of parameter normalization. Thus: $R = r/a, u = U/U_0, T = Dt/a^2$.

It will be seen that the separation of variables for the cylindrical case is not so straightforward. The derivation of a solution leads to Bessel functions.

8.4 Problem 5.4 was concerned with the construction of a one-dimensional CA based on the operation described in eqn (5.8), namely:

$$_{k+1}C(x) = 1 \quad \text{if } {}_kC(x-1) + {}_kC(x+1) \text{ is odd}$$
$$_{k+1}C(x) = 0 \quad \text{if } {}_kC(x-1) + {}_kC(x+1) \text{ is even}.$$

Devise an identical 10 node CA where node 1 and node 10 are fixed at value 1 at all times. The initial population of the remaining nodes should be an arbitrary binary sequence. The value of a particular node (e.g. node 5) should be sampled over a number of iterations and then subject to a DFT to determine the spectral content of the CA to investigate whether it is a function of starting population. The process could also be repeated as time progresses to see whether the spectral content is dependent on the overall number of iterations.

Would it be possible to use this as a method of quantifying the state of order of the system?

8.5 Fig. 1.2 (reproduced in Fig. 8.16) was concerned with the aiming sights for an airborn free-fall bomb. Some of the same concepts can be applied to synthetic aperture radar (SAR).

The distance between an airborn radar and some target on the ground below can be given in terms of the height of the aircraft and the range.

$$r = \sqrt{h^2 + d^2} \quad \text{or} \quad r = h\left[1 + \frac{d^2}{2h^2} - \frac{d^4}{8h^2} + \dots\right].$$

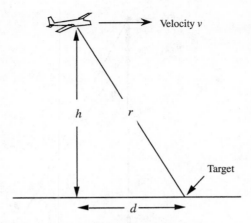

Fig. 8.16 The model for synthethic aperture radar.

So long as the target range d is kept small with respect to the height h then we can write:

$$r = h + \frac{d^2}{2h}.$$

Now the radar signal is transmitted at a single frequency. The returned signal displays several properties. There are phase shifts at the transmitted frequency and there are Doppler shifts. The phase shift due to a target at d is given by:

$$\phi(d) = -\left[\frac{2\pi}{\lambda}\right]2r.$$

Show that the phase shift is quadratically dependent on the range.
    The Doppler history can be given in terms of phase as:

$$\frac{1}{2\pi}\frac{d}{dt}\phi(d).$$

Use this to develop an expression for the Doppler frequency in terms of height, range, and aircraft velocity.

8.6 Let us suppose that the aircraft described in the previous problem is fitted with one transmitter and two receivers which are set one above the other and separated by a distance B, see Fig. 8.17. The two receivers will have slightly different phase information for each pixel of target surface. The phase difference between the two receivers can be expressed as

$$\phi = \frac{2\pi}{\lambda}(r_2 - r_1).$$

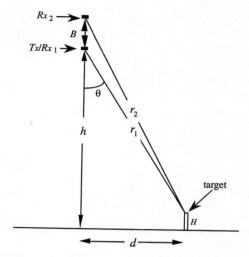

Fig. 8.17 The SAR model with one transmitter and two receivers.

The cosine rule can be used to replace r_2 and thus in terms of the geometry:

$$\phi = \frac{2\pi}{\lambda}\left(\sqrt{r_1{}^2 + B^2 + 2r_1B\cos\theta} - r_1\right)$$

In normal circumstances ϕ, the phase difference between the two received signals, is measured and recorded. Since r_1, h, B, and λ are known you should use this information to develop expressions for r_2 and θ and hence deduce and expression for the height of a target which is given by:

$$H = h - r_1 \cos\theta.$$

It may be of interest to note that if measurements are recorded using a single receiver at two different times and if the data can be registered pixel by pixel, then the phase difference can yield information about topographical changes (either buildings removed or constructed or indeed minor changes due to earthquake upheavals).

8.7 The following exercise may be difficult to undertake, but the outcome could be very interesting. The instantaneous voltages across the two outputs of a Loasby voice-synthesis network could be summed as follows:

$$V_T = a\,V_{nasal} + (1-a)V_{mouth}.$$

This value could be stored in computer memory and the binary equivalent, padded out to the appropriate byte size could then be fed to a digital to analogue converter (such as a computer sound blaster) for acoustic output. The objective would then be to investigate the effects of line impedances, output impedances (oral and nasal), as well as resonator characteristics on the perceived frequency when the overall 'filter' has been applied to white noise generated in the trachea.

The next step would be to investigate the effects of trachea start-up transients on a time-invariant filter.

Finally, you could combine a time-dependent response of the trachea with time-dependent changes in aesophagus characteristics in order to generate isolated sounds. Normal joined speech would presumably involve time-dependent changes in oral and nasal resonances.

Some additional techniques

This book has provided only a brief introduction to numerical modelling of physical problems relevant to engineering. It would be quite impossible to provide a detailed treatment of all topics within a reasonably sized text. Our objective has been to help the reader to identify classes of problems, and the varieties of treatments available for their solution. To this end we have always attempted to provide relevant and accessible references at the end of each chapter.

We have emphasized from the start the importance of a coherent planned approach and have suggested five basic steps in the evolution of a successful model:

(1) problem specification;

(2) problem simplification;

(3) model specification;

(4) model solution;

(5) model verification.

We would now like to identify some techniques which have not so far been mentioned. In the sections below we provide thumb-nail sketches of four important methods with references for further study.

9.1 Green's function methods

Many of the concepts which have been presented in this book in differential form could have been presented in integral form. Although this is generally considered a more difficult approach, it has certain advantages which we will touch on in this section. Green's theorem states that the integral of the curl of a vector field over the surface is equal to the closed integral of the field along the line which bounds the surface. Using this we can present Poisson's equation $\Delta^2 \phi = -\rho/\varepsilon_r \varepsilon_0$ in an integral form as:

$$\int \int \int \frac{\rho(r')\mathrm{d}^3 r}{4\pi \varepsilon_r \varepsilon_0 |r - r'|} = \phi(r). \tag{9.1}$$

Thus the charge density is on the left-hand side of the equation and the potential at each point $\phi(r)$ is on the right. Here $\phi(r)$ is known and $\rho(r')$ is to be solved; the reverse of the situation in the differential representation. The equation depicts a convolution of the charge at every position r' as it affects point r. The Green's function which describes the effect at r due to charges at r' embodies this information. It will be seen that the Green's function approach is a natural consequence of some of the numerical treatments in earlier chapters.

Example 9.1 Green's function approach to heat-flow modelling
We start with a conventional heat-flow equation:

$$\frac{\partial^2 T(x,t)}{\partial x^2} + g(x,t) = \frac{1}{D}\frac{\partial T(x,t)}{\partial t}$$

where $T(x,0) = F(x)$.

The function $F(x)$ represents an initial distribution of temperature and $g(x, t)$ is the contribution from some heat generator at $t > 0$.
A solution for $T(x, t)$ is given by:

$$T(x,t) = \int_{x'=-\infty}^{\infty} G(x,t:x',0)F(x')\mathrm{d}x' + D\int_{\tau=0}^{t}\int_{-\infty}^{\infty} G(x,t:x',\tau)g(x',\tau)\mathrm{d}x'\mathrm{d}\tau.$$

This might look awesome but it is in fact elegant. It states that the problem can be represented as the superposition of two independent problems, one being the temperature/time distribution due to an initial temperature profile, the other being due to $g(x, t)$.

If we take each of these contributors separately we can get a slightly better understanding of the nature of G. Let us first assume that there is no heating source, so that the value of $T(x, t)$ depends only on an initial input $F(x)$, which is defined only for the point x_0. The use of the Dirac delta function helps here. The initial condition $T(x, 0) = F(x)$ can be redefined as:

$$F(x) = F'_0\, \delta(x - x_0)$$

so $F(x)$ is zero everywhere except at x_0 where it is F'_0 times the delta function. The subsequent temperature profile can then be given as:

$$T(x,t) = F'_0\, G(x,t:x_0,0)$$

Now let us assume that the initial temperature is everywhere zero so that the profile is due only to one generator at position x_0 and time t_0. We can write

$$g(x,t) = \Delta T\, \delta(x - x_0)\, \delta(t - t_0)$$

and
$$T(x,t) = \Delta T\, G(x,t\!:\!x_0,t_0).$$

It is not always easy to define G analytically, particularly in problems with realistic boundaries, although the book by Beck *et al.* [9.1] is a good introduction on the application to heat-flow problems. Thus before we complete this example by indicating how a discrete approach might be used we will first revisit some material from Section 5.1.1.

Discrete Green's functions

In the context of material which has been presented in Chapters 3, 4, and 5 it is quite easy to define a discrete Green's function [9.2]. Starting with the equation for a two-dimensional simple random walk:

$$\begin{aligned}
{}_{k+1}N(x,y) = \frac{1}{4}\{ &{}_kN(x-1,y) + {}_kN(x+1,y) \\
&+ {}_kN(x,y-1) + {}_kN(x,y+1)\}
\end{aligned} \tag{9.2}$$

we can apply this formula to monitor what we call number diffusion. This is shown in Fig. 9.1 during the first few iterations of a single-shot input of magnitude 1024 injected at $x = 0, y = 0$ at $t = 0$. The number 1024 was chosen merely to avoid the generation of decimals by division during 5 time steps.

If we treat the numbers in the figure as the concentrations of a diffusing species then, as in Section 5.1.1, they follow the Bernoulli trial:

- for $|x| + |y| > k$ ${}_kC(x,y) = 0$;
- for $k - x - y$ odd ${}_kC(x,y) = 0$;
- for $|x| + |y| \le k$ ${}_kC(x,y)$ is given by eqn (9.9);
- for $k - x - y$ even ${}_kC(x,y)$ is given by eqn (9.9).

So we have

$$_kC(x,y) = {}_0C(x_0,y_0)\frac{1}{4^k}\begin{bmatrix} k \\ \frac{k-x+y}{2} \end{bmatrix}\begin{bmatrix} k \\ \frac{k-x-y}{2} \end{bmatrix}. \tag{9.3}$$

Returning to the example we should be able to see that eqn (9.3) is identical to the equation $T(x,t) = F_0' = G(x,t\!:\!x_0,0)$. Therefore, we are able to define the discrete Green's function for this part of the problem as:

$$G(x,t\!:\!x_0,0) = \frac{1}{4^k}\begin{bmatrix} k \\ \frac{k-x+y}{2} \end{bmatrix}\begin{bmatrix} k \\ \frac{k-x-y}{2} \end{bmatrix}.$$

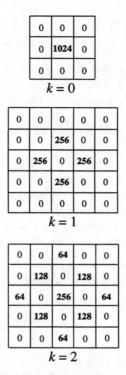

Fig. 9.1 Number diffusion during the first two time steps.

9.2 The method of moments

The method of moments is a *residual* technique for solving systems of equations. Readers who have studied Chapter 7 will see much in common with popular techniques in FEM. In this section we follow the presentation of Harrington [9.3].

In a one-dimensional representation of Poisson's equation

$$\frac{\partial^2 V(x)}{\partial x^2} = \frac{\rho(x)}{\varepsilon_r \varepsilon_0} \tag{9.4}$$

we can identify the following components:

$\dfrac{\partial^2}{\partial x^2}$ the operator is designated as L

$V(x)$ the response is designated as f

$\dfrac{\rho(x)}{\varepsilon_r \varepsilon_0}$ the source is designated as g

Equation (9.4) can now be restated as:

$$Lf = g \qquad (9.5)$$

We can now say that if $Lf = g$ exists and is unique, then there exists an inverse operator L^{-1} such that $f = L^{-1}g$. But, you might ask, is this anything more than straight integration? The answer would be a guarded *yes* as will be seen below.

The first step is to expand f in a series of functions

$$f = \sum_n \alpha_n f_n \qquad (9.6)$$

where the α_n values are constants and f_n are expansion or basis functions. If the summation is to infinity then the substitution of eqn (9.13) into eqn (9.12) will lead to an exact solution, otherwise it will lead to an approximate solution. This can be presented in terms of a running example.

Example 9.2 Solution of a differential equation using moment methods
Solve

$$-\frac{\partial^2 V(x)}{\partial x^2} = 1 + 4x^2$$

within the limits $V(0)$ and $V(1) = 0$.
Since the source term $(1 + 4x^2)$ has the characteristics of a power series, there will be a similar expansion for the response:

$$f_n = x - x^{n+1}$$

so that

$$f = \sum_n^N \alpha_n(x - x^{n+1}).$$

Our objective is then to determine the values of α_n and this is where we must temporarily leave this example and return to some apparently heavy mathematics.

For the example, we specifically need an inner product $<f, g>$, whose definition properties are discussed in detail by Harrington [9.3]. Since the limits of our example are $0 \leq x \leq 1$ a suitable product is

$$<f, g> = \int_0^1 f(x)g(x)\mathrm{d}x. \qquad (9.7)$$

We can also define an inner product $< Lf, g >$ which is required in order to determine the adjoint operator, L^a which is defined as follows:

$$< Lf, g > = < f, L^a g > .\qquad(9.8)$$

Using our definitions above we can show that

$$\int_0^1 \frac{\partial^2 f(x)}{\partial x^2} g(x)\, dx = \int_0^1 f(x) \frac{\partial^2 g(x)}{\partial x^2}\, dx$$

So that $L^a g = \dfrac{\partial^2 g}{\partial x^2}$ and thus in the example $L^a = L$.

Equation (9.7) is not unique. It can be shown that another, equally satisfactory inner product would be:

$$< f, g > = \int_0^1 w(x) f(x) g(x)\, dx\qquad(9.9)$$

Where $w(x)$ is a weighting or test function. This is similar to the method of weighted residuals in FEM. This weight has the property:

$$\sum_n \alpha_n < w_m, Lf_n > = < w_m, g > \quad \text{where } m = 1, 2, 3, \ldots.\qquad(9.10)$$

All of this can be presented in matrix form as:

$$\boldsymbol{L\alpha = g}\qquad(9.11)$$

$$\text{where } \boldsymbol{L} = \begin{bmatrix} < w_1, Lf_1 > & < w_1, Lf_2 > & \cdots \\ < w_2, Lf_1 > & \vdots & \\ \vdots & & \end{bmatrix}, \boldsymbol{\alpha} = \begin{bmatrix} \alpha_1 \\ \alpha_2 \\ \alpha_3 \\ \vdots \end{bmatrix},$$

$$\text{and } \boldsymbol{g} = \begin{bmatrix} < w_1, g > \\ < w_2, g > \\ < w_3, g > \\ \vdots \end{bmatrix}.$$

If $\boldsymbol{L}$ is nonsingular then an inverse $\boldsymbol{L^{-1}}$ exists so that

$$\boldsymbol{\alpha = L^{-1} g}.\qquad(9.12)$$

This means that f can now be determined as soon as we know $w(x)$. In the chapter on FEM, we had to determine ϕ and we mentioned there that there were several possible approaches. In the Galerkin method $\phi_n = f_n$. In the present context an equivalent approach ($w_n = f_n$) is quite satisfactory.

Example 9.2 (continued)

Returning to our example, l_{mn}, the (m, n)th element of $L = \langle w_m, Lf_n \rangle$ can be expressed as:

$$\int_0^1 (x - x^{m+1})\frac{\partial^2}{\partial x^2}(x - x^{n+1})dx = \frac{m\,n}{m+n+1}$$

similarly, g_m, the mth component of $g = \langle w_m, g \rangle$ is given by:

$$\int_0^1 (x - x^{m+1})(1 - 4x^2)dx = \frac{m(3m+8)}{2(m+2)(m+4)}.$$

Now, the sum required is from $n = 1$ to N. If $N = 1$ then $l_{11} = 1/3$, $g_1 = 11/30$, so that $\alpha_1 = 11/10$, which leads to:

$$f = \frac{11}{10}\left[x - x^2\right].$$

If $N = 2$ then the equation reduces to

$$\begin{bmatrix} \frac{1}{3} & \frac{1}{2} \\ \frac{1}{2} & \frac{4}{5} \end{bmatrix}\begin{bmatrix} \alpha_1 \\ \alpha_2 \end{bmatrix} = \begin{bmatrix} \frac{1}{30} \\ \frac{7}{12} \end{bmatrix}$$

so that $\alpha_1 = \frac{1}{10}$ and $\alpha_2 = \frac{2}{3}$.

From this we obtain $f = \frac{1}{10}\left[x - x^2\right] + \frac{2}{3}\left[x - x^3\right]$. which is a much better approximation to the analytical solution.

If we wish to increase the accuracy even further, we can set $N = 3$ to get $\alpha_1 = \frac{1}{2}$, $\alpha_2 = 0$, and $\alpha_3 = \frac{1}{3}$ producing

$$f = \frac{1}{2}\left[x - x^2\right] + \frac{1}{3}\left[x - x^4\right].$$

This is an exact solution and is obtained for any higher value of N.

The example given was tractable, but it is not always possible to integrate analytically in order to find L. In other cases it may be necessary to use numerical integration, or to force the solution to satisfy the original equation only at discrete points within its range. These aspects in the context of electromagnetic applications are discussed extensively by Harrington [9.3].

Sadiku [9.4] covers the application of moment methods for the derivation of Green's functions in electromagnetics.

9.3 The boundary-element method

Whereas the finite-element method described the responses of subdivisions of a region, the boundary-element method (BEM) subdivides the boundary of the region with the solutions taken as a combination of the exact solutions inside the region. Beer and Watson [9.5] outline the associated steps as follows:

- Discretize the boundary (only the boundary). For a plane, this is equivalent to a set of one-dimensional finite elements. For a solid, planar boundary elements are used.

- Apply a variational method to the unknowns inside each boundary element. This is identical to finite-element methods, but only on the boundary.

- Use a fundamental solution of the governing differential equation which satisfies the governing differential equation exactly.

- Calculate contributions to the coefficient matrices.

- Assemble element contributions.

- Solve the system of equations using matrix methods. This step refers only to the boundary.

- Use the boundary information to compute values inside the domain.

Discrete element modelling

This is a related technique which has been pioneered by Biçanic and colleagues [9.6]. It uses an approach similar to FEM/BEM applied to a many-bodied system whose components are called discrete elements, and considers in particular the interaction of these elements at their boundaries. It is an ideal technique for simulating material fragmentation such as the breaking of a window due to impact, the response of steel to an armour piercing shell, and rock trajectories during quarry blasting.

9.4 Spectral-domain methods

We end this chapter with a brief mention of an important method which solves spatial problems by means of a transformation into the frequency domain in order to overcome numerical difficulties. The outline given below is due to Scott [9.7] and uses as its running example the response of a monochromatic electromagnetic wave impinging on a very thin metallic sheet of infinite surface area.

An incident electric field induces currents in the surface which in turn creates a scattered field which radiates from the surface. In vector terms this scattered component can be expressed as:

$$^sE(r) = -j\omega\mu \left[A(r) + \frac{1}{k^2} \nabla \left[\nabla \cdot A(r) \right] \right]. \tag{9.13}$$

The incident field vector can be resolved into Cartesian components, $^iE_x(x,y)$ and $^iE_y(x,y)$. Equation (9.13) can also be expressed in Cartesian components as:

$$^sE_x(x,y) = -j\omega\mu \left[A_x(x,y) + \frac{1}{k^2} \left[\frac{\partial^2 A_x(x,y)}{\partial x^2} + \frac{\partial^2 A_y(x,y)}{\partial x \partial y} \right] \right]$$
$$^sE_y(x,y) = -j\omega\mu \left[A_y(x,y) + \frac{1}{k^2} \left[\frac{\partial^2 A_x(x,y)}{\partial x \partial y} + \frac{\partial^2 A_y(x,y)}{\partial y^2} \right] \right]. \tag{9.14}$$

The vector magnetic potential $A(r)$ is related to the vector current density J in the surface through the Green's function:

$$A(r) = \int_{\text{sources}} \int G(r:r') J(r') \, dS'. \tag{9.15}$$

The vector Green's function in this type of problem is given by:

$$G(r:r') = \frac{e^{jk|r-r'|}}{4\pi|r-r'|} \tag{9.16}$$

where k is the wave number $2\pi/\lambda$. On the surface of the metallic sheet the total tangential electric field is zero, so that $^iE_{\text{tangential}} + {}^sE_{\text{tangential}} = 0$. Therefore, we can write:

$$-{}^iE(r) = -j\omega\mu \left[A(r) + \frac{1}{k^2} \nabla \left[\nabla \cdot A(r) \right] \right]. \tag{9.17}$$

This electric field integral equation can be solved, but there are considerable problems in the vicinity of the sources themselves, when $r = r'$. These problems can be circumvented by transforming into the spectral domain where eqn (9.17) can be solved over a single cell if the reflecting surface is periodic.

The Cartesian components of incident electrical field vector $^iE(x,y)$ can be expressed in terms of wave or spectral properties:

$$^iE_x(x,y) = {}^iE_x(\alpha_0, \beta_0) \, e^{j(\alpha_0 x + \beta_0 y)}$$
$$^iE_y(x,y) = {}^iE_y(\alpha_0, \beta_0) \, e^{j(\alpha_0 x + \beta_0 y)} \tag{9.18}$$

where $\alpha_0 = k \sin(^i\theta) \cos(^i\phi)$, $\beta_0 = k \sin(^i\theta) \sin(^i\phi)$.

The magnetic field vector, $A(x,y,z)$ in Cartesian coordinates can be expressed in the spectral domain by the Fourier integral

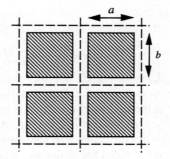

Fig. 9.2 Section of a periodic conducting surface.

$$A(x,y,z) = \int\limits_{-\infty}^{\infty} \int\limits_{-\infty}^{\infty} A(\alpha,\beta)_{z=0} e^{j(\alpha x + \beta y)} e^{\pm jz\sqrt{k^2 - \alpha^2 - \beta^2}} \, d\alpha \, d\beta. \tag{9.19}$$

If $k^2 > \alpha^2 + \beta^2$ then we have a plane propagating wave.

If $k^2 < \alpha^2 + \beta^2$ then we have a decaying or evanescent wave.

If the reflecting surface is periodic as shown in Fig. 9.2, then the currents and fields in one cell can be defined and then related to the same parameters in other cells by a phase shift.

The surface current within a cell is given by:

$$J(x+a,y+b) = J(x,y) \, e^{j(\alpha_0 x + \beta_0 y)} \, e^{2\pi m j} e^{2\pi n j} \tag{9.20}$$

where $m = 1, 2, \ldots$ and $n = 1, 2, \ldots$. This means that only discrete values of α and β in eqn (9.19) are allowed, so that

$$\alpha_m = \alpha_0 + \frac{2\pi m}{a} \quad \text{and} \quad \beta_n = \beta_0 + \frac{2\pi n}{b}.$$

Equation (9.19) then reduces to a Fourier series

$$A(x,y,z) = \sum_m \sum_n A(\alpha_m, \beta_n)_{z=0} \, e^{j(\alpha_m x + \beta_n y)} \, e^{\pm jz\sqrt{k^2 - \alpha_m^2 - \beta_n^2}}. \tag{9.21}$$

The individual elements of this series are known as *Floquet harmonics*. This can now be substituted into eqn (9.14) to obtain a matrix expression for the scattered field components:

$$\begin{bmatrix} {}^s E_x(\alpha_{m'}\beta_n) \\ {}^s E_y(\alpha_{m'}\beta_n) \end{bmatrix} = \begin{bmatrix} 1 - \frac{\alpha_m^2}{k^2} & -\frac{\alpha_m \beta_n}{k^2} \\ -\frac{\alpha_m \beta_n}{k^2} & 1 - \frac{\beta_n^2}{k^2} \end{bmatrix} \begin{bmatrix} A_x(\alpha_{m'}\beta_n) \\ A_y(\alpha_{m'}\beta_n) \end{bmatrix}. \tag{9.22}$$

Remembering that $A(\alpha_m, \beta_n)$ can be expressed as a Green's function, we can therefore Fourier transform eqn (9.21) to yield:

$$A(\alpha_{m'} \beta_n) = G(\alpha_{m'} \beta_n) J(\alpha_{m'} \beta_n) \tag{9.23}$$

where $G(\alpha_m, \beta_n)$ is the transform of the scalar Green's function $G(x, y)$. This is of major importance: in the spatial domain $A(x, y)$ was expressible as a convolution, in the frequency domain it is merely the multiplication of spectral quantities. In our example, the spectrum of $G(\alpha_m, \beta_n)$ is continuous and it is $J(\alpha_m \beta_n)$ which has the discretized spectrum. Thus by operating in the spectral domain we have avoided the difficulties which ocurred at $r = r'$ in the spatial domain. This convenience is the justification of the spectral domain method.

Returning to the solution of the example problem, it can be shown [9.6] that the Fourier transform of $G(x, y)$ is given by:

$$G(\alpha_{m'} \beta_n) = \frac{-j}{2\sqrt{k^2 - \alpha_m^2 - \beta_n^2}} \quad \text{for } k^2 > \alpha^2 + \beta^2. \tag{9.24}$$

If this and eqn (9.23) are substituted into eqn (9.22) we obtain:

$$\begin{bmatrix} {}^s E_x(\alpha_{m'} \beta_n) \\ {}^s E_y(\alpha_{m'} \beta_n) \end{bmatrix}_{z=0} = \frac{-1}{2we} \begin{bmatrix} \dfrac{k^2 - \alpha_m^2}{\sqrt{k^2 - \alpha_m^2 - \beta_n^2}} & -\dfrac{\alpha_m \beta_n}{\sqrt{k^2 - \alpha_m^2 - \beta_n^2}} \\ -\dfrac{\alpha_m \beta_n}{\sqrt{k^2 - \alpha_m^2 - \beta_n^2}} & \dfrac{k^2 - \beta_n^2}{\sqrt{k^2 - \alpha_m^2 - \beta_n^2}} \end{bmatrix} \begin{bmatrix} Jx(\alpha_{m'} \beta_n) \\ Jy(\alpha_{m'} \beta_n) \end{bmatrix}.$$

The spectrum of E is known because the spectrum of J is known. Therefore the above equation can be expressed as the sum of the $n \times m$ contributions to represent the electric field equation (eqn (9.24)) in the spectral domain, which although it applies to all the conducting components of the surface, needs to be analysed over a single cell only.

The spectral-domain technique can be used to great benefit in any modelling situation which treats the interaction of waves with a periodic scattering surface. It can also be extended to aperiodic and singular structures.

References

9.1 J.V. Beck, K.D. Cole, A. Haji-Sheikh and B. Litkouhi, *Heat conduction using Green's functions*, Hemisphere Publishing Corp. 1992.

9.2 D. de Cogan and P. Enders, Discrete Green's functions and hybrid modelling of thermal and particle diffusion. *International Journal of Numerical Modelling* **7** (1994) 407–418.

9.3 R.F. Harrington, *Field computation by moment methods*, Robert E. Krieger Publishing Co, Malabar, Florida 1983.

9.4 M.N.O. Sadiku *Numerical techniques in electromagnetics*, CRC 1992, pp. 315–334.

9.5 G. Beer and J.O. Watson, *Introduction to finite and boundary element methods for engineers*, John Wiley and Sons, Chichester 1994.

9.6 See several relevant papers in Proc. 4th Int. Conf. on Non-linear Engineering, Swansea 1991, Pineridge Press.

9.7 C. Scott, *The spectral domain method in electromagnetics*, Artech House, 1989.

Appendix: Mathematical fundamentals

This appendix presents skeleton details of various mathematical tools which are used throughout the book. The intention is to provide a quick *aide-memoir* for those who may have forgotten and a brief introduction (just to get started) for those for whom some of the ideas may be relatively new. The objective throughout is to outline the techniques within the context of their use in the various chapters and in so far as is possible sources of reference for further study are provided.

A.1 Coordinate systems and transformations

There are three widely used methods of defining a point in space. They are:

(a) the Cartesian coordinate system;

(b) the cylindrical polar coordinate system;

(c) the spherical polar coordinate system.

The Cartesian coordinate system uses three mutually orthogonal axes (x, y, and z) to determine the position of the point P (see Fig. A.1).

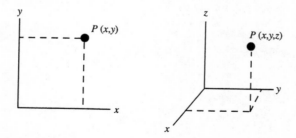

Fig. A.1 The definition of a point P in two- and three-dimensional Cartesian coordinates.

The cylindrical polar coordinate system uses r, the distance from the origin to the point P, and the angle θ, which this line subtends from the horizontal axis, as a means of location in two dimensions. In three dimensions the height h of the point above the horizontal plane is also specified.

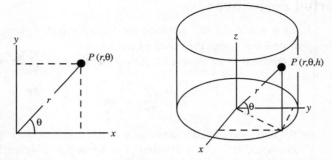

Fig. A.2 The definition of a point P in two- and three-dimensional cylindrical coordinates.

It is easy to see from Fig. A.2 that the following relationships provide a means of transforming between the cylindrical and Cartesian systems:

$$r = \sqrt{x^2 + y^2} \qquad \theta = \tan^{-1}\left[\frac{y}{x}\right] \qquad h = z$$
$$x = r \cos \theta \qquad y = r \sin \theta \qquad z = h. \tag{A.1}$$

The spherical polar coordinate system uses one distance and two angles as shown in Fig. A.3.

The transformations between Cartesian and spherical polar coordinates are as follows:

$$r = \sqrt{x^2 + y^2 + z^2} \qquad \phi = \tan^{-1}(y/x) \qquad \theta = \tan^{-1}\left[\frac{z}{\sqrt{x^2 + y^2}}\right] \tag{A.2}$$
$$x = r \cos \phi \sin \theta \qquad y = r \sin \theta \sin \theta \qquad z = r \cos \theta.$$

Some further examples of transformations will be given in the section on matrices below.

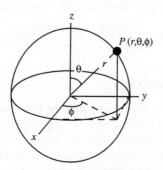

Fig. A.3 The definition of a point P in spherical polar coordinates.

A.2 Partial derivatives

Section 3.3 in the text outlined the basis for partial derivatives. This section will deal only with the practical aspects.

Suppose that we have a function $Z(x, y) = 3x^3y^2 + 2x + 4y$. We can define partial derivatives:

$$Z_x = \frac{\partial Z}{\partial x} \quad \text{and} \quad Z_y = \frac{\partial Z}{\partial y}. \tag{A.3}$$

In the first case y is considered as a constant so that we rewrite the function as $Z(x, y) = Ax^3 + 2x + B$ where $A = 3y^2$ and $B = 4y$ are both constants. The derivative Z_x is then

$$Z_x = 3Ax^2 + 2 \quad \text{or} \quad Z_x = 9y^2x^2 + 2.$$

Similarly $Z_y = 6x^3y + 4$.

The second derivatives present a higher level of complexity, because we can now define:

$$Z_{xx} = \frac{\partial^2 Z}{\partial x^2}, \ Z_{yy} = \frac{\partial^2 Z}{\partial y^2}; \quad \text{and} \quad Z_{xy} = \frac{\partial^2 Z}{\partial x \partial y}, \ Z_{yx} = \frac{\partial^2 Z}{\partial y \partial x}. \tag{A.4}$$

It is worthwhile to check whether $Z_{xy} = Z_{yx}$ and to calculate the value of $Z_{xx} + Z_{yy}$. It would also be useful to consult Kreyszig [A.1] for details of partial derivatives of vector functions (grad, div, and curl) and their transformations from Cartesian to cylindrical and polar coordinate systems.

If we have a function $W = F(x, y)$ where x $X(t)$ and $y = Y(t)$ we can define a total derivative:

$$\begin{aligned} dW &= \frac{\partial F}{\partial x} dx + \frac{\partial F}{\partial y} dy \\ &= \frac{\partial F}{\partial x} \frac{\partial x}{\partial t} dt + \frac{\partial F}{\partial y} \frac{\partial y}{\partial t} dt \end{aligned} \tag{A.5}$$

so that the partial derivative of W with respect to t is given by:

$$\frac{\partial W}{\partial t} = \frac{\partial F}{\partial x} \frac{\partial x}{\partial t} + \frac{\partial F}{\partial y} \frac{\partial y}{\partial t}. \tag{A.6}$$

A.3 Analytical methods

This section is concerned with the basic properties of two power series. Taylor series is first encountered in Chapter 3 where it provides an approximation in finite-difference formulations. It can also be used to determine the order of accuracy for a particular discretization. Maclaurin series provide very useful expansions for a range of well-known functions and are used in Chapter 4.

If x_0 is chosen as reference and if we measure the distance from this point as $x - x_0$ then a power series in $x - x_0$ has the form:

$$a_0 + a_1(x - x_0) + a_2(x - .x_0)^2 + \ldots \tag{A.7}$$

where a_0, a_1, a_2, etc. are coefficients and x is a variable. If this series converges to some value $f(x)$ then we can write that:

$$f(x) = a_0 + a_1(x - x_0) + a_2(x - x_0)^2 + \ldots . \tag{A.8}$$

In Taylor series the values of the coefficients can be determined at $x = x_0$ by successively differentiating the function. This yields:

$$f(x) = f(x_0) + (x - x_0)\frac{df(x_0)}{dx} + \frac{(x - x_0)^2}{2!}\frac{d^2f(x_0)}{dx^2} + \ldots . \tag{A.9}$$

As in Chapter 3, we can consider a differential equation of the form

$$\frac{dy}{dx} = f(x, y).$$

Expressing the finite interval (Δx) as 'h' we can present this as a Taylor series:

$$y(x + h) = y(x) + h\,y'(x) + \frac{h^2}{2!}\,y''(x) + \ldots$$

which can be rewritten as:

$$y(x + h) = y(x) + h\,f(x, y) + \frac{h^2}{2!}\,f'(x, y) + \ldots .$$

If h is sufficiently small that h^2 and higher terms can be ignored then:

$$y(x + h) = y(x) + h\,f(x, y)$$

which is the fundamental justification for the Euler–Cauchy method.

If the reference point $x_0 = 0$ then we have a Maclaurin series:

$$f(x) = f(0) + x\frac{df(0)}{dx} + \frac{x^2}{2!}\frac{d^2f(0)}{dx^2} + \ldots . \tag{A.10}$$

If $f(x) = \cos(x)$ then the Maclaurin series expansion is:

$$\cos(x) = \cos(0) + x[-\sin(0)] + \frac{x^2}{2!}[-\cos(0)] + \frac{x^3}{3!}[\sin(0)] + \frac{x^4}{4!}[\cos(0)] + \ldots$$

$$= 1 + 0 - \frac{x^2}{2!} + 0 + \frac{x^4}{4!} + \ldots .$$

Similarly, it can be shown that:

$$\sin(x) = x - \frac{x^3}{3!} + \frac{x^5}{5!} - \frac{x^7}{7!} + \cdots$$

and that:

$$e^x = 1 + x + \frac{x^2}{2!} + \frac{x^3}{3!} + \cdots .$$

A.4 Matrix methods

Matrices are mentioned at various places throughout this text and consist of a rectangular array of m rows and n columns represented by:

$$A = \begin{bmatrix} a_{11} & a_{12} & a_{13} & \cdots & a_{1n} \\ a_{21} & a_{22} & \cdots & & \\ a_{31} & \cdots & & & \\ \vdots & & & & \\ a_{m1} & & & & a_{mn} \end{bmatrix}.$$

The terms a_{11}, a_{22}, .. are termed the diagonal elements. Matrices can be treated using the conventional arithmetic operators:

$$\text{if } B = \begin{bmatrix} a & b \\ c & d \end{bmatrix} \text{ and if } Z = \begin{bmatrix} v & w \\ x & y \end{bmatrix} \text{ then } B \pm Z = \begin{bmatrix} a \pm v & b \pm w \\ c \pm x & d \pm y \end{bmatrix}.$$

Matrices can be multiplied by scalars:

$$2\begin{bmatrix} 1 & 7 \\ 5 & 3 \end{bmatrix} = \begin{bmatrix} 2 & 14 \\ 10 & 6 \end{bmatrix}.$$

Multiplication of a matrix by a matrix is more complicated and requires care. Using the definitions of B and Z above

$$BZ = \begin{bmatrix} av + bx & aw + by \\ cv + dx & cw + dy \end{bmatrix}.$$

This is quite different from

$$ZB = \begin{bmatrix} av + cw & bv + dw \\ ax + cy & bx + dy \end{bmatrix}.$$

Matrix multiplication is not commutative.

Matrix multiplication can be used as a short-hand method for expressing systems of linear equations and applications can be seen in Section 3.3.1 and is used extensively in Chapter 7. For instance a shape function from finite-element analysis can be expressed as:

$$V_1 = a + bx_1 + cy_1$$
$$V_2 = a + bx_2 + cy_2$$
$$V_3 = a + bx_3 + cy_3$$

or as

$$\begin{bmatrix} V_1 \\ V_2 \\ V_3 \end{bmatrix} = \begin{bmatrix} 1 & x_1 & y_1 \\ 1 & x_2 & y_2 \\ 1 & x_3 & y_3 \end{bmatrix} \begin{bmatrix} a \\ b \\ c \end{bmatrix}.$$

There are some other, very useful definitions in matrices. The transpose of a matrix swaps rows and columns so that

$$\boldsymbol{B}^{\mathrm{T}} = \begin{bmatrix} a & c \\ b & d \end{bmatrix}.$$

The determinant of matrix $\boldsymbol{B}$ is given by $|\boldsymbol{B}| = ad - cb$. The determinant of
$\begin{bmatrix} a & b & c \\ d & e & f \\ g & h & i \end{bmatrix}$ is

$$\begin{vmatrix} a & b & c \\ d & e & f \\ g & h & i \end{vmatrix} = a \begin{vmatrix} e & f \\ h & i \end{vmatrix} - b \begin{vmatrix} d & f \\ g & i \end{vmatrix} + c \begin{vmatrix} d & e \\ g & h \end{vmatrix}$$

which can be further simplified to $a(ei - hf) - b(di - gf) + c(dh - ge)$. Determinants have many applications, but are particularly important for finding the inverse of a matrix. The inverse of $\boldsymbol{B}$ is defined so that

$$\boldsymbol{B}\boldsymbol{B}^{-1} = \begin{bmatrix} 1 & 0 \\ 0 & 1 \end{bmatrix}$$

which is called the unit matrix I. There are a variety of methods for finding matrix inverses and these are covered in Deif [A.2]. The inverse is used in the solution of linear equations, e.g. an electrical network might be expressed by the following equation

$$\begin{bmatrix} V_1 \\ V_2 \\ V_3 \end{bmatrix} = \begin{bmatrix} R_1 + R_2 & -R_2 & 0 \\ -R1 & R_2 + R_3 + R_4 & -R_3 \\ 0 & -R_3 & R_4 + R_5 \end{bmatrix} \begin{bmatrix} I_1 \\ I_2 \\ I_3 \end{bmatrix}$$

$$\mathbf{V} = \mathbf{RI}$$

where the components of the voltage matrix are known. The components of the current matrix can be determined by:

$$I = R^{-1}V .$$

Gaussian elimination

There may be situations where the use of inverse methods for solution might be inappropriate and instead it might be possible to use an alternative method which has much in common with conventional methods for solving simultaneous equations. We can use the previous example (with some values inserted) to demonstrate the concept of Gaussian elimination:

$$\begin{bmatrix} -2 \\ 3 \\ 12 \end{bmatrix} = \begin{bmatrix} 4 & -3 & 0 \\ -3 & 10 & -5 \\ 0 & -5 & 6 \end{bmatrix} \begin{bmatrix} I_1 \\ I_2 \\ I_3 \end{bmatrix} \text{ can be written as } \begin{bmatrix} 4 & -3 & 0 & -2 \\ -3 & 10 & -5 & 3 \\ 0 & -5 & 6 & 12 \end{bmatrix}$$

The a_{11} term (4) can be treated as the pivot so that if 3 times the first row is added to 4 times the second row we get:

$$\begin{bmatrix} 4 & -3 & 0 & -2 \\ 0 & 31 & -20 & 6 \\ 0 & -5 & 6 & 12 \end{bmatrix}.$$

The 31 in the middle row becomes the pivot so that if we add a factor of 31/5 of this row to the row below we will obtain:

$$\begin{bmatrix} 4 & -3 & 0 & -2 \\ 0 & 31 & -20 & 6 \\ 0 & 0 & 86 & 402 \end{bmatrix}.$$

This third row now states that $86 I_3 = 402$. I_3 is therefore determined. This is then back-substituted into the line above to yield I_2 since $3 I_2 - 20 I_3 = 6$. Further back-substitution into $4 I_1 - 3 I_2 = -2$ gives I_1. The following set of Maple instructions use slightly different pivots, but back-substitution gives the same results.

```
a:= matrix (3,4,[4,−3,0,−2,−3,10,−5,3,0,−5,6,12]);
with(linalg):
gausselim(a);
```

Gauss–Seidel is another matrix method for solving systems of linear equations, but as it involves iterative methods it is discussed in Section A.6.

A.5 Discrete methods

This section starts with the subject of interpolation over equidistant intervals and then goes on to consider aspects of discrete calculus.

Interpolation

Simple interpolation attempts to estimate the value of a function at x given data at x_0 and x_1 where $x_0 < x < x_1$

$$f(x) = f_0 + (x - x_0) \left[\frac{f_1 - f_0}{x_1 - x_0} \right]$$

(where $x_1 = x_0 + h$).

There are times when this does not give sufficient accuracy and we will now consider interpolation using forward, backward, and central difference schemes.

The first forward difference at x_j is $\Delta f_j = f_{j+1} - f_j$.

The second forward difference at x_j is $\Delta^2 x_j = \Delta f_{j+1} - \Delta f_j$.

In similar fashion to the Maclaurin series expansion of a function we can define the Newton–Gregory forward difference interpolation as:

$$f(x) = f_0 + r \Delta f_0 + \frac{r(r-1)}{2!} \Delta^2 f_0 + \dots$$

where

$$r = \frac{x - x_0}{h}.$$

Although it is normal to define a derivative as $\dfrac{f(x + \Delta x) - f(x)}{\Delta x}$ there is no special reason why it should not be defined as $\dfrac{f(x) - f(x - \Delta x)}{\Delta x}$.

In a similar way we can define backward differences:

The first backward difference at x_j is $\nabla f_{j,} = f_j - f_{j-1}$.

The kth backward difference at x_j is $\nabla^k f_j = \nabla^{k-1} f_j - \nabla^{k-1} f_{j-1}$.

We can similarly define the Newton–Gregory backward difference interpolation as

$$f(x) = f_0 + r \nabla f_0 + \frac{r(r+1)}{2!} \nabla^2 f_0 + \dots$$

where

$$r = \frac{x - x_0}{h}.$$

We can also have similar definitions using central differences:

The kth central difference at x_j is $\delta^k f_j = \delta^{k-1} f_{j+1/2} - \delta^{k-1} f_{j-1/2}$

The accuracy of an interpolation does not always increase as the order of the interpolation used; numerical instability can become a factor. The alternative method of splines is discussed by Kreyszig [A.1].

Numerical integration

There are various techniques for obtaining the value of an integral; the Monte Carlo method was mentioned in Section 6.9 (as a student of chemistry one of the authors used to cut out the region under a curve and compare its mass with that of a known area of the same graph paper). This section outlines the two best-known methods.

If we take the function xe^{-x} we could approximate it by a series of rectangles as shown in Fig. A.4. In each case x^* is the midpoint which intersects the function. If the range of integration is between a and b and if we define as:

$$h = \frac{(b-a)}{n}$$ where n is the number of rectangles then:

$$\int_a^b f(x) \doteq h[f(x_1^*) + f(x_2^*) + f(x_3^*) + \ldots]$$ called the *rectangular rule*.

It can be seen that this is not going to be a particularly good approach since the approximation has not accurately accounted for the limits.

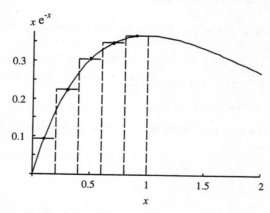

Fig. A.4 A representation of the function xe^{-x} in the range $0 \leq x \leq 2$, with a piecewise approximation using rectangles in the range $0 \leq x \leq 1$.

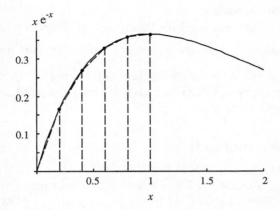

Fig. A.5 The function xe^{-x} in the range $0 \leq x \leq 2$, with a linear approximation using chords of the curve in the range $0 \leq x \leq 1$.

We can do much better by using a piecewise linear approximation of the function using polygons (Fig. A.5). In this case the integral is given by:

$$\int_a^b f(x) \doteq h[\frac{f(a)}{2} + f(x_1) + f(x_2) + \ldots f(x_{n-1}) + \frac{f(b)}{2}]$$ called the *trapezoidal rule*.

We can demonstrate this for $n = 10$ using Maple.

```
f:= x*exp(−x);
with(student);
trapezoid(f,x=0..1,10);
evalf(value("));
```

The answer, 0.2634, compares quite well with the analytical result, 0.2642.

Simpson's rule for numerical integration uses a piecewise quadratic approximation. The interval h is now defined as:

$$h = \frac{(b-a)}{2n}.$$

If $x_0 = a$ and $x_{2n} = b$ then the integral can be approximated by:

$$\frac{h}{3}[(f(x_0) + 4f(x_1) + 2f(x_2) + 4f(x_3) + \ldots + 2f(x_{2n-2}) + 4f(x_{2n-1}) + 2f(x_{2n})].$$

The application of Simpson's rule to the running example gives a result which is identical with the analytical result to four decimal places.

Numerical differentiation

This appeared implicitly throughout the chapter on finite-difference methods. However, there are times when the definition $\lim\limits_{h \to 0} \dfrac{f(x+h) - f(x)}{h}$ does not provide a sufficiently accurate description. In this situation it may be necessary to differentiate some suitable polynomial. Some examples are given in Kreyszig [A.1].

A.6 Iterative methods

Fixed-point iteration

The equation $x = 2^{-x} + 1$ seems at first sight almost impossible to solve, but is a prime candidate for treatment using a fixed-point iterative scheme. This expresses the equation in the following form:

$$x_{k+1} = 2^{-x_k} + 1$$

The initiation of this sequence may need some trial and error; an unsuitable value may result in a divergence of the iteration, but an initial value of $x_0 = 0.5$ is suitable and converges fairly rapidly:

$$
\begin{aligned}
x_1 &= 2^{-0.500} + 1 &&= 1.707 \\
x_2 &= 2^{-1.707} + 1 &&= 1.306 \\
x_3 &= 2^{-1.306} + 1 &&= 1.404 \\
x_4 &= 2^{-1.404} + 1 &&= 1.3777 \\
x_5 &= 2^{-1.3777} + 1 &&= 1.3848 \\
x_6 &= 2^{-1.3848} + 1 &&= 1.3829 \\
x_7 &= 2^{-1.3829} + 1 &&= 1.3834 \\
x_8 &= 2^{-1.3834} + 1 &&= 1.3833 \\
x_9 &= 2^{-1.3833} + 1 &&= 1.3833397
\end{aligned}
$$

and so on.

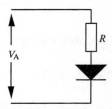

Fig. A.6 Diode/resistor circuit for a fixed-point iterative scheme.

There are times when a fixed point iteration scheme will diverge if the equation is expressed in one form, but will converge if the equation is rearranged. The following simple example from electronics demonstrates this point. The diode in Fig. A.6 is driven from a fixed source V_A through a resistor R and it is required to estimate the voltage V_R across the resistor.

This voltage is given by:

$$V_R = IR$$

but the current is controlled by the diode, i.e.

$$I = I_0 \left[e^{q(V_A - V_R)/kT} - 1 \right]$$

where I_0 is the leakage current and q is the electron charge. Thus

$$V_R = RI_0 \left[e^{q(V_A - V_R)/kT} - 1 \right].$$

If reasonable values are chosen (e.g. $V_A = 1$ V, $R = 10\,\Omega$, $I_0 = 10^{-9}$A and $q/kT = 0.05$ V) for the constants, then we have the following scheme:

$$_{k+1}V_R = 10^{-8} \left[e^{q(V_A -_k V_R)/kT} - 1 \right].$$

We can start the iterative process with $_0V_R = 0$ and we will then get:

$$_1V_R = 10^{-8} \left[e^{20} - 1 \right] = 4.85 \text{ V i.e. greater than } V_A!$$

Other initial choices of $_0V_R$ might give a better first result but in general, unless the choice is very lucky the iterative process will diverge.

On the other hand the diode equation could have been rearranged as:

$$I = \frac{V_R}{R} \quad \text{and} \quad \frac{I - I_0}{I_0} = e^{q(V_A - V_R)/kT}$$

so that

$$\frac{kT}{q} \left[\log_e \frac{I - I_0}{I_0} \right] = V_A - V_R$$

or

$$V_R = V_A - \frac{kT}{q} \left[\log_e \frac{I - I_0}{I_0} \right].$$

We can now express an iterative scheme in terms of current:

$$_{k+1}I = \frac{1}{R} \left[V_A - \frac{kT}{q} \left[\log_e \frac{I - I_0}{I_0} \right] \right].$$

This converges; I can be found and therefore V_R can be found. Thus, we can conclude that it is the model and not the circuit that goes out of control if wrongly manipulated.

Newton–Raphson iteration

This scheme attempts to find the root x_0 of an equation on the basis that the addition of a small correction h will fulfil the equation. i.e. $f(x_0 + h) = 0$.

This can be expressed by a Taylor series and to a first approximation we can say:

$$f(x_0 + h) = f(x_0) + h \frac{df(x_0)}{dx} = 0$$

or

$$h = \frac{-f(x_0)}{\left[\dfrac{df(x_0)}{dx} \right]}.$$

If this is viewed as a first estimate so that $x_1 = x_0 + h_0$ then we can calculate a new estimate of the correction

$$h_1 = \frac{-f(x_1)}{\left[\dfrac{df(x_1)}{dx} \right]},$$

and thereafter $x_2 = x_1 + h_1$.

We could attempt to find a solution of

$$f(x) = \cos(x) - x = 0.$$

Now $\dfrac{df}{dx} = -\sin(x) - 1$

therefore

$$h = \frac{\cos(x) - x}{-\sin(x) - 1}$$

which yields

$$x_{n+1} = x_n + \frac{\cos(x) - x}{\sin(x) + 1}.$$

Starting with $x_0 = 0.5$ the results at successive iterations are:

$x_1 = 0.75522$

$x_2 = 0.73914$

$x_3 = 0.73908$

$x_4 = 0.73908$

$x_5 = 0.73908.$

Secant method

This is identical in form to the Newton method except that the derivative is replaced by the finite difference form i.e.

$$x_{n+1} = x_n - \frac{f(x_n)}{\left[\dfrac{df(x_1)}{dx}\right]}$$

where

$$\frac{df(x_n)}{dx} = \frac{f(x_n) - f(x_{n-1})}{x_n - x_{n-1}}.$$

Gauss–Seidel iteration

This is an iterative technique which finds solutions for a system of linear equations. It is particularly valuable for dealing with implicit schemes such as those we encountered in the finite-difference chapter. Let us imagine that we have the following set of equations where the only knowns are $T(0) = 20$ and $T(5) = 100$

$$T(1) = \frac{T(0) + T(2)}{2}$$

$$T(2) = \frac{T(1) + T(3)}{2}$$

$$T(3) = \frac{T(2) + T(4)}{2}$$

$$T(4) = \frac{T(3) + T(5)}{2}.$$

These can be rearranged as:

$$
\begin{aligned}
T(1) &= & T(2)/2 & & & + & & 10 \\
T(2) &= & T(1)/2 & + & T(3)/2 & & & \\
T(3) &= & T(2)/2 & & & + & T(4)/2 & \\
T(4) &= & & & T(3)/2 & + & & 50 \,.
\end{aligned}
$$

The terms on the right of each equation are treated as 'known' quantities in an iterative scheme while the terms on the left are the 'new' values, so that the previous set of equations read as:

$$
\begin{aligned}
_{n+1}T(1) &= & _nT(2)/2 & & & + & & 10 \\
_{n+1}T(2) &= & _nT(1)/2 & + & _nT(3)/2 & & & \\
_{n+1}T(3) &= & _nT(2)/2 & & & + & _nT(4)/2 & \\
_{n+1}T(4) &= & & & _nT(3)/2 & + & & 50 \,.
\end{aligned}
$$

We can start the iterations with a set of guesses (for the purpose of example, we have chosen a particularly bad set $_0T(1) = 90$, $_0T(2) = 50$ $_0T(3) = 40$, $_0T(4) = 30$, so that we get:

$$
\begin{aligned}
_1T(1) = & \quad 50/2 + 10 \quad = 35 \\
_1T(2) = & \quad 90/2 + 40/2 = 65 \\
_1T(3) = & \quad 50/2 + 30/2 = 45 \\
_1T(4) = & \quad 40/2 + 50 \quad = 70 \ .
\end{aligned}
$$

If this process is repeated enough times then the results will be: $T(1) = 36$, $T(2) = 52$, $T(3) = 68$, $T(4) = 84$ which are identical with the analytical result. This is the *Jacobi method* of iterative solution.

The process can be speeded up if each value is immediately substituted into the remaining equations within the set immediately it has been calculated and this is called the *Gauss–Seidel method*. In terms of simple BASIC code this might read as:

```
FOR ITERATION = 1 TO 20
   NEW1 = TOLD2/2 + 10
      TOLD1 = TNEW1
   TNEW2 = TOLD1/2 + TOLD3/2
      TOLD2 = TNEW2
   TNEW3 = TOLD2/2 + TOLD4/2
      TOLD3 = TNEW3
   TNEW4 = TOLD3/2 + 50
      TOLD4 = TNEW4
NEXT ITERATION
```

References

A.1   E. Kreyszig, *Advanced engineering mathematics* (6th Edition), John Wiley and Sons 1988.
A.2   A.S. Deif, *Advanced matrix theory for scientists and engineers,* Abacus Press (Gordon and Breach Science Publishers) 1991.

Index